HOW TO BE A
CLIMATE OPTIMIST

ALSO BY CHRIS TURNER

Planet Simpson
The Geography of Hope
The Leap
The War on Science
How to Breathe Underwater
The Patch

HOW TO BE A
CLIMATE
OPTIMIST

BLUEPRINTS FOR
A BETTER WORLD

CHRIS TURNER

RANDOM HOUSE CANADA

PUBLISHED BY RANDOM HOUSE CANADA

www.penguinrandomhouse.ca

Library and Archives Canada Cataloguing in Publication

Title: How to be a climate optimist : blueprints for a better world / Chris Turner.
Names: Turner, Chris, 1973- author.
Identifiers: Canadiana (print) 20210273178 | Canadiana (ebook) 20210273968 |
    ISBN 9780735281974 (softcover) | ISBN 9780735281981 (EPUB)
Subjects: LCSH: Sustainability. | LCSH: Sustainable living. | LCSH: Climate
    change mitigation.
Classification: LCC GE196 .T87 2022 | DDC 304.2—dc23

Text design: Matthew Flute
Cover design: Leah Springate
Image credits: (Blurred railway track and green bush) baona/iStock/Getty Images

MIX
Paper from
responsible sources
FSC® C016245
FSC www.fsc.org

Printed in Canada

10 9 8 7 6 5 4 3 2 1

Penguin
Random House
RANDOM HOUSE CANADA

*For Sloan, who teaches me every day how
to get better at building a better world.*

*And for Alexander, who reminds me every day
why that better world is so urgently needed.*

"We inhabit, in ordinary daylight, a future that was unimaginably dark a few decades ago, when people found the end of the world easier to envision than the impending changes in everyday roles, thoughts, practices that not even the wildest science fiction anticipated. Perhaps we should not have adjusted to it so easily. It would be better if we were astonished every day."

—Rebecca Solnit, *Hope in the Dark*

CONTENTS

3.0 THE MUCH BETTER DECADE

4.0 THE AGE OF TRANSITION

1.0

HOW TO BE A
CLIMATE OPTIMIST

1.1 A Better World Waits

I'm a climate optimist. There's nothing starry-eyed or Pollyanna-like about it. It's not a slogan or a marketing pitch. It's something I earned. By accident, in a sense.

By the time these lines meet a reader's eyes, my oldest child will be seventeen years old. I first became obsessed with the scale of the mounting climate catastrophe in the year before we started our family, horrified by the coming devastation. This was not a common sentiment in those days. From the point of view of someone worrying over the real issues of the day (war, terrorism, corporate malfeasance, the economy, the economy, the economy), climate fears were fringe concerns to be found over a horizon too distant to even see. But I'd talked to a few climate scientists and read some reports in the obscure technical science news, and I was convinced to my core that there would be no public issue in my lifetime as consequential as our collective response to climate change. And I wondered how to tell this horror story with enough shock and awe to make it as urgent as a dispatch from the front lines in a war zone. I outlined a series of magazine features, a sort of world tour of the world's climate hot spots—disappearing Pacific islands, melting glaciers, receding coastlines, dried-up farmlands. A map of catastrophe. I had an interested editor ready to write a cheque to cover the round-the-world ticket I would need to write those stories. And then the magazine he edited went bankrupt, and the climate horror stories shifted to my mental back burner as I started a family.

With my wife pregnant with our first child and the horizon of my responsibility stretching beyond my own lifetime, I reacquainted myself with my shelved climate-hot-spots idea. But then I imagined my kid's earliest memories perhaps being coloured by trips to islands that would become uninhabitable and glaciers soon to vanish forever. What a grotesque gift that would be. And so I stood the idea on its head. What would a world tour of places that were *beating* climate change look like?

I'd never given focused thought to what sustainability or resilience means, to how energy is made at national or global scales, to what causes a society's (or a world's) population to choose to burn fossil fuels to meet nearly all its daily needs. But I had a notion—that someone already had to be working on solutions. That those solutions might work. That I might be able to stitch them together into a makeshift map of a better world. That became my primary occupation for the next seventeen years (and counting). I measured the depth of my experience on the beat by the birthdays and milestones of my kids. I watched the oldest one march off to a student climate strike not too long ago, feeling not guilt at what the world had become or terror at what the future might bring, but instead confidence in the certainty—cautious and qualified, but a certainty nonetheless—that the tools to build that better world were waiting for those kids. This is what I mean when I say I've earned my optimism.

If I had little to offer my children after seventeen years of searching, I wouldn't be able to make any reasonable claim to optimism. If I'd scoured the world and interviewed the experts and attended all the wonky conferences and trade shows and come away with nothing to show for it but best intentions and wishful thinking, I'd be delusional, if not downright fraudulent, in my optimism. Thankfully, my naive hunt has returned an abundance. I've seen the solutions, tracked their development and growth, watched them spread from the freak fringe of garage tinkerers and hardcore green survivalists and—okay, yes—Scandinavians to the front-page news of big business, global politics, mainstream industry and everyday life.

I've been to the better world. I've talked to its architects and toured its facilities. I've seen the tools that work and the ones that don't. When I say I'm a climate optimist, I mean that, in the face of the existential challenge of climate change, I believe the world will develop and implement better systems and technologies to meet our daily needs and reduce our global greenhouse gas emissions to somewhere very near zero in this century. It likely won't happen in time to halt the permanent alteration of many of the planet's ecosystems and prevent significant dislocation and suffering for millions of people, but it will occur with enough speed and thoroughness to leave a foundation for civilization durable enough for future generations to build their dreams on. And that's about the same as any generation's ever done for the ones to come.

I'd like to show you how I became a climate optimist. Set aside your anxiety and dread and skepticism for a little while—they'll

still be there when I'm done. Let's see where a little hope and in-
genuity and even some exuberance and delight can take us.

1.2 Doom's Limits

Doom is easy now.

Maybe doom has always been low-effort, but it's straight-up
effortless now. Catastrophe, Armageddon, nuclear winter or zombie
apocalypse—these scenes are everywhere because they're not hard
to dream up. No need even to invest your imagination in the work
of inventing nightmare images and fantastical tales of roaring
hellfire or wrath-of-god hurricanes or swelling pestilence—they
are recurring features in the nightly news and the stuff of memes
up and down every social media feed. The sky is on fire, the oceans
in tumult, the glaciers vanishing, the fish stocks collapsing, the
coral reefs bleached ghostly white. The end is surely nigh, or at
least one of the more likely nighs.

Doom is easy indeed. We're steeped in doom, awash in devastation, swimming in despair. And our brains are always up for more,
riddled with biases and tics that make us hyper-aware of impending
threats and prone to assume that one freak misfortune will soon
lead to more of the same. That's why everyone still wants a ticket
to the disaster movie. Everyone still sings along to the sad songs of
collapse. *A hard rain's a-gonna fall. It's the end of the world as we
know it. This is it, the apocalypse.* Buzzfeed posts a listicle under the
title "17 Signs the World Is Definitely Over by Next Thursday,
Latest"—everyone clicks. But then Thursday comes and goes and
there's another day and another after that. That's when you need
something more than impending doom to keep you going.

So let's set doom aside. Not because the end is any less nigh, necessarily, not because the catastrophe is averted by denying its existence, but because we have to keep going in the face of it. Let's focus instead on the bed we have to lie down in regardless, the home it's in, the way the lights stay on and the place is kept warm or cool, how we get from there to work and school and back again. Let's focus on living. Living better.

1.3 Embrace Dark Euphoria

Bruce Sterling is a science-fiction author and thinker who has spent his career examining the ways rapidly advancing technology is reshaping human existence. He is perhaps best known as one of the progenitors of the sci-fi genre known as cyberpunk, which emerged in the early 1980s and anticipated the staggering wave of innovations and transformations in digital technology that are now ubiquitous. Sterling is interested not so much in what technology can do as in how it shapes the culture—how it makes people feel and act, how it defines their sense of what is possible and desirable (and what is horrifying). Back in 2009, at a progressive tech conference called Reboot 11 in Copenhagen, he delivered a closing keynote address on the subject of "grand future narratives."

"What is the cultural temperament of this era?" Sterling asked the small knot of hackers and coders in the room. "Well, I think it's got a good two-word summary: *Dark euphoria*. Dark euphoria is what the twenty-teens feels like. Things are just falling apart, you can't believe the possibilities. It's like anything is possible, but you never realized you're going to have to dread it so much. It's like a leap into the unknown."

Sterling's dark euphoria meme never really caught on—not like the Roaring Twenties, the Lost Generation, "turn on, tune in, drop out," the "Me" Decade, the Slacker Generation. Maybe it's one of those coinages that looks most prophetic in retrospect, or perhaps it'll simply flow away forever down the noisy currents of the internet. Which would be unfortunate, because it so succinctly captures the general mood of these first decades of the twenty-first century.

The darkness arises from a mounting sense that the concept of human progress that has driven idealists and inspired huddled masses for centuries has stalled out. That promise of progress, in Sterling's summary, was this: "You get more scientific knowledge, you create more tools, make more jobs, you master nature, you get more power, cheaper power, you struggle for a better life for your children, you're looking for health, prosperity, material security, shelter, bigger, faster, stronger, knowing more. Everybody knows that's progress." A brighter future, basically, or at least one with more opportunity and wonder in it than this one. And that vision of progress appears to have vanished. "The actual objective situation looks more like this: No money, scarcity, financial collapse, collapsed states, general precarity, an energy crisis, low-intensity global warfare, and a rapidly advancing climate crisis."

On the climate front, Sterling rails against what he calls "hair-shirt green" environmentalism. For many of us, this is what we think of as environmental activism's (and thus climate activism's) default mode. Do less. Consume less. Live in smaller apartments, drive smaller vehicles, take shorter trips and fewer flights. Switch to reusable shopping bags, outlaw plastic straws. Maybe put off having a family, a child. Opt out of the joy of creation. To which Sterling responds, "Stop acting dead." Hair-shirt green, he says, is simply a reversal of polarity on the existing system, an inversion, a series of rejections. It's not creative or inventive, let alone exciting or inspiring. It's an itchy shirt, there to remind you constantly that you use too much fabric.

I didn't discover Sterling's dark euphoria talk until ten years after he first delivered it, but I recognized it immediately as a concise description of the vibe I'd spent that decade immersed in. The bright ideas and exciting breakthroughs, as well as the painstaking, unglamorous work of reimagining a better world to replace the dark one. I spent the decade hunting for dark euphoria. Looking for the cracks in everything, to borrow Leonard Cohen's indelible line, because that's how the light gets in.

1.4 An Age of Offhand Miracles

The fuel of my climate optimism, the reliable spark that keeps it moving, is a phenomenon I've come to think of as an offhand miracle.

That's how I caught the bug. Way back in 2000, I was sent to Montreal by *Time* magazine to report on environmentally friendly cars. The occasion was one of those somewhat obscure trade shows that would soon become central to my journalistic beat, an event called the 17th International Electric Vehicle Symposium—EVS-17.

Given what's happened to electric vehicles in the twenty years since, I should set the stage here. As of 2000, nearly all the world's major automakers had begun to tinker with the elusive promise of the emissions-free car, electric or otherwise. Toyota had the first version of the Prius there, and Nissan had a prototype of the Leaf. There were those little Smart cars alongside glorified golf carts, motorbikes and scooters. But this whole showcase was a minor sidelight on the heart of the automotive business. The major car companies had brought their prototypes and test models out to a racetrack in Montreal for us all to take for a test spin, but nobody was in any hurry to get those vehicles to a show-room near you. (Even the Prius as you now know it wouldn't make it to North American Toyota dealers until 2003.)

The real star of EVS-17 wasn't even a plug-in electric or hybrid. It was Ford's P2000, an experimental sedan that ran on hydrogen gas and spat potable water and nothing else from its tailpipe. That was the car that attracted the crowd of journalists and motoring geeks at the racetrack, and it's the one that I featured most prom-inently in my *Time* story. Because a fully functional Ford sedan whose only exhaust is water is pretty goddamn miraculous. And it was offhand, too, in the way that a test-lab engineer from Ford showing you the droplets of water on his hand as he squats next to a tailpipe on a racetrack at an obscure technical conference is intrinsically offhand. It launched me on a quest—twenty years and counting—to find other offhand miracles.

That quest has been more fruitful than anyone's wildest dreams that day in Montreal. Because here's the thing: Everyone thought that day there would be Priuses and Leafs on the road before too long, maybe some little all-electric shoebox made in Europe for crowded city streets or a future car-share program. And some-day someone might crack the range problem of fully electric cars and/or the cost problem of hydrogen-fuelled vehicles, and then those would find a place on a few roads away from a test track too. Twenty years from then? Maybe thirty? Sure, it could happen.

But no one had the slightest notion, not the vaguest inkling or the wildest premonition, that what would actually happen was some weird, megalomaniacal engineer who was working at the time on the thorny problem of small-scale internet business transactions at PayPal would cash out a couple of years later, use his dot-com riches to take over a fledgling manufacturer of experimental electric cars, and turn that company into not only the first successful automotive start-up in America in more than half a century but also the maker of the first mass-market all-electric sports car in history—all in barely more than a decade. But that's what Elon Musk did with Tesla Motors. A fantastical idea too wildly speculative and distant-future-tense to even make it to a racetrack in Montreal in 2000 is now everyday reality. Keep that in mind as we discuss the limits of what's possible for the next ten or twenty years.

Now, if the miraculous rise of Tesla were the only story of its genre, the only wild arc from impossible fantasy to everyday reality I'd discovered in twenty years on this climate-solutions beat, I would have a hard time holding it up as the avatar of anything. Happily, that's not the case.

My first focused reporting on this beat was a visit I made in 2005 to a Danish island called Samsø, which was endeavouring to become the world's first island powered entirely by renewable energy. Fifteen years later, not only has Samsø surpassed its own goal, it is a pacesetter for the entire country and all of the European Union, which has pledged to duplicate the island's achievement by 2050. (Denmark itself intends to eliminate all emissions from its electricity grid by 2030.) When I first began reporting on solar power in 2005, there were five gigawatts' worth of photovoltaic panels connected to all the electricity grids on earth, and it was common sense to suggest that solar would never amount to more than one percent of the world's electricity supply. It was at 0.2 percent at the time. In 2020, China alone connected almost 50 gigawatts of new solar power to its grids, the largest share of 127 gigawatts added worldwide that year, for a grand total of 707 gigawatts—more than 3 percent of global production and growing at a pace considered sheer fantasy back in 2005.

The first bike-share system I ever encountered was a novelty at the Copenhagen train station, and I rode my rented bike to the harbour and back on the first physically separated bike lane I'd ever seen. Bike and scooter shares are now commonplace in hundreds of cities around the world, including my own, where I can ride downtown and back on 10 kilometres of real separated bike lanes.

I went to the desert outside Taos, New Mexico, in 2006 to inspect a crazy hippie fever dream called an Earthship—a house designed to use the natural heating and cooling of the sun to achieve self-sufficiency for its energy needs. There is a now an eight-storey apartment building in downtown Vancouver that uses "passive house" design principles to do roughly the same thing. By 2032, every new building in the province of British Columbia will be built to a code broadly consistent with those "net zero" standards.

I tracked a series of these victorious arcs in their paths from margin to mainstream. Offhand miracles, one after another.

As an Aside: On Terminology

I'll need to use shorthand terms from here on for the cumulative wave of offhand miracles I'm describing. There are a number of them out there in the world already, phrases and descriptions all gathering adherents and inspiring memes and getting clumsily overused at board meetings. There are *green energy* and simply *going green*; *clean energy* and *clean technology* and *cleantech*; *eco-friendly* and *climate-friendly* and *net zero*; *carbon-free, low-carbon economy* and *emissions-free energy*; *sustainability* and *resilience* and *adaptation*. Some of these have precise technical meanings in some fields, but all have been muddied and watered down to some degree. I will use some of these terms in contexts where they are the most useful way to describe a thing, but I don't want to get lost in the weeds or stuck in some ideological corner with whatever baggage those terms sometimes carry.

So I'll ask for your indulgence of some oversimplification. When I want to refer succinctly to a tool (a technology, policy, concept, idea, gadget or notion) that is part of a general effort to reduce the amount of catastrophic climate change humanity is likely to cause, I will call such a tool a "climate solution." When I'm referring specifically to the uneven collective worldwide effort to implement a range of these climate solutions to create a world as near to entirely free of greenhouse gas emissions as quickly as possible, I will call it the "global energy transition" (or simply "energy transition").

Cool? Cool.

1.5 Less Bad and Much Better

The Tesla Model S, the company's first mass-market sedan, was not designed to be gasoline-free, by which I mean that was not its primary goal. The intention was not to eliminate the greenhouse gas emissions from a standard car. No, the idea was to make the best car on the road, which merely happened to be all-electric as well. And so when the Model S hit the streets, it was not just a car you fuelled through an electrical cord but also the fastest car of its type, the smartest car of its type, the safest car on the road. It reportedly broke the machine used to test a vehicle's response to a rollover accident—the kind of next-level safety feature that Volvo built its entire brand around, simply another detail on a Tesla.

To ride in a Tesla when I first did, back in 2014, was to suddenly find yourself inside a World's Fair exhibit of futuristic wonders as the humdrum world of outmoded Toronto whizzed by. I remember being particularly taken by the all-electric acceleration. The motors in electric cars don't need to rev up like some tired old internal combustion engine—they feed maximum power to the drivetrain all at once. A Tesla S pulling out at an intersection pushes you back, gently but firmly, into the passenger seat. It reminded me of one of those spinning centrifugal-force rides on a midway, the floor falling away as gravity holds the riders aloft against the wall. More than that, though, it reminded me of a train ride I took through Spain in 2009.

The train is called the AVE del Sol. AVE is short for *alta velocidad española* ("Spanish high speed"). And it's called *del sol* ("of the sun") because it carries passengers back and forth along a high-speed rail line between the Spanish capital of Madrid and Malaga, the largest city on Spain's beach-studded Costa del Sol.

I boarded just before noon, and by lunchtime I was hurtling across the Spanish countryside at 302 kilometres per hour, by far the fastest I'd ever travelled over the surface of the earth and easily the fastest any normal civilian has been able to go in all of human history. And I was at complete ease. I had a cold Andalusian sherry in my hand and lunch on its way, and I was reading a book in total comfort. Your car does not ride this smoothly at city driving speeds, I can assure you. I would arrive at my destination, the centre of Madrid, an hour after the lunch dishes were cleared, a 540-kilometre journey completed in less than three hours. And I wouldn't have to wait at a baggage carousel or find a taxi to the city centre. I'd have no need of parking. With that sherry in hand, the country racing past, and a digital display tracking the novelty of our speed, I was quite certain I was participating in the absolute pinnacle achievement to date in the annals of human transport. This wasn't simply a faster train. It wasn't just a marginal improvement over a six-hour highway drive or the grind of airport security. It was *much better*.

In retrospect, that train ride to Madrid was my pivot point. I'd been reporting on climate solutions for five years by then, but my focus had been mostly on how one vehicle produced fewer emissions than another, how one power plant wasn't as dirty as another, how one building needed less energy to heat and cool than another. I'd been looking at how to make the existing fabric of modern industrial life *less bad*. In the years after that Spanish high-speed train ride—a decade that saw the fastest growth of a new global energy system since the rapid expansion of the fossil fuel industry in the late 1800s—I began to seek not just ways to slow down climate change but also tools for a more desirable and more durable way of life. The target of my pursuit shifted forever from *less bad* to *much better*.

I'm not big on celebrating technology for its own sake. My first beat as a journalist, before I moved on to climate solutions, was reporting on the first giddy wave of the internet boom, and I barely grasped the big-picture point of half of what I reported on. Sure, you could do all your grocery shopping on the internet—but, really, *why?* What was so bad about going to the store? I'm typing this on an eight-year-old laptop, which almost makes it an antique in the digital age. You get my point. I don't really care how fast a Tesla Model S can go from zero to sixty (the updated 2016 Model S with the "ludicrous mode" option does it in 2.5 seconds, if you must know). But the importance of the AVE trip from Malaga to Madrid is not about technology. It's about living better. It's about taking the extraordinary power, knowledge and sophistication of twenty-first-century science and engineering and sociopolitical coordination and using them to their fullest, vastly improving the experience of getting from one place to another. And doing so not for some elevated elite, but as the backbone of a national transportation network.

I want to unpack this fully. It needs extra emphasis. We are entirely too inured to sudden dramatic technological enhancements, bored by the staggering power of our gadgets moments after we upgrade them. We don't stop and simply stare in awe often enough at the offhand miracles.

For most of human existence, transportation was wearying, perilous, a necessary evil—an ordeal to be endured. Prior to 1800, most people had no regular transport options other than to walk, slowly, often through dangerous terrain. The epic storytelling of Chaucer's *Canterbury Tales* transpires as a party of pilgrims entertain themselves on the multi-day walk from London to Canterbury Cathedral, a journey of barely 100 kilometres. A skilled rider on horseback could cover more ground—60 kilometres per day, on average—and a chariot on good Roman roads was known to cover 100 kilometres in a day. Boats moved slightly faster, perhaps as fast as 20 kilometres per hour with a full sailing rig, a strong wind and a team of oarsmen. Still, the vast majority of people spent their entire lives barely ever straying farther than a day's walk from their homes.

Such was the pace and scale of life in much of the world when James Watt's ingenious new steam engine was first harnessed for experiments in locomotion in the early 1800s. By the end of the century—almost overnight by historical standards—steel tracks were crisscrossing the continents and trains were routinely roaring along them at 100 kilometres per hour. (As early as the 1930s there were locomotives capable of reaching speeds of 200 kilometres per hour.) But that turned out to be only the second most important transportation advance of the age. Here was the real revolutionary step: take a small engine, mount it in a horseless carriage, and feed it the energy-dense petroleum first discovered in large quantities in the 1860s—suddenly the ease and speed of the divine rested in the hands of any individual of means. And then the Ford Motor Company began rolling Model T cars off its assembly line in 1908, and the locomotive power of the internal combustion engine was within reach for those of even fairly modest means.

With the rise of the automobile as a primary form of transportation and, in time, the reconfiguration of whole cities to its specifications—a shift that was neither an accident nor a mistake at the time—people discovered real mobility, escape, freedom, seemingly infinite possibility. Cities were dirty, smog-choked, crowded and dangerous—and now you could live at their leafy edges and make a Canterbury pilgrimage's worth of travel to and from work every day in effortless comfort. Thus was born the suburban ideal, a wild new concept for daily life in which the nuclear families of the rapidly expanding middle class could live each on their own private estate on a quiet street with a car (or two! or more!) in the garage. Errands run by car, school mornings and weekend sports by car, using the lavish amenities of the new communities that ringed the cities, holidays down wide highways anywhere the family wanted to go—all at the pace of a speeding locomotive. It was, in the sales language native to the scene, an exceedingly enticing value proposition.

Today, talking as we are about the existential threat of climate change, this might well seem like recklessness or pure madness— or corporate malfeasance, mass manipulation, straight-up consumerist propaganda. But at first, when this value proposition fully emerged at the end of a terrifying war, predicated on the promise of a new era of peace and prosperity for all, it was simply *much better*. In fact, this vision of daily life remains so seductive that, as I write, you can find it fully intact on Tesla's website—a glossy photo illustration of an expansive, elegant suburban home powered by the sun, with ample battery storage in the garage for all that solar power and a Tesla charging up in the driveway.

Which brings me back to Tesla and the Spanish train and the hidden cost of that twentieth-century value proposition. We should look on that fossil-fuelled suburban ideal as an understandable but regrettable error, a Faustian bargain with a half-century payback. In the name of mobility and prosperity, many of us bought into a deal that slowly, steadily exhausted the planet's ability to stay healthy enough to sustain us. Transportation alone is the source of 14 percent of global greenhouse gas emissions, and it's one slice of an entire way of life built on easy mobility and cheap fuel that forms the core of the climate crisis. Which is why an electric car is far from the whole solution.

We need to reinvent the whole value proposition. And we have.

1.6 A Much Better World

I embarked on this journey in search of climate solutions almost by accident, but I continued along it because I hoped to witness the dawn of a new age. It was a speculative quest, and it felt by turns impossible and then inevitable and then impossible again. It was the only way out, but it felt throughout like a road that would not be taken. And now a global energy transition is under way, moving faster by the day, and its progress is inevitable. When it's over, renewable energy will be ubiquitous and fossil fuels rare. If greenhouse gas emissions never quite reach zero, they will be reduced near enough to zero to ensure that whatever ails the planet will no longer be getting worse.

I find myself returning again and again to a photo I found a while back that accompanied a story about high-speed trains. It looks like one of those illustrations designers put together to show what a much better future might look like if all the climate solutions I've spent years researching come to fruition. It depicts a landscape with an elevated rail line bisecting it, a sleek white bullet train racing along the track. To the left of the track in the foreground is a dense knot of mid-rise apartment buildings, like a little square of European cityscape dropped into the middle of the countryside. To the right of the track, a tidy cross-hatch of farm fields adds bright splashes of green and gold to the picture. There are low green hills in the background, and I've seen enough hill country crowned with wind turbines to have no trouble at all imagining they're there, providing clean power to the transportation network. There is something simultaneously timeless and futuristic about the image, a sense of durable urban civilization merged seamlessly with rural sustenance.

The image is not an artist's rendering of some grand plan or a computer-generated collage of future-tense living. It's a snapshot of the Xi'an-to-Chengdu high-speed rail line, which runs through Shaanxi province in northwestern China. Xi'an is an ancient city, the eastern starting point of the Silk Road, a travel hub since before the birth of Christ on a trade route that stretches from central China all the way to the Mediterranean Sea. This is what gives that photo its power: it's real, up and running, an entirely functional twenty-first-century scene. That stretch of track is one small piece of a 36,000-kilometre network of Chinese high-speed rail lines— enough transport infrastructure to stretch from sea to sea across North America and back, six times over. And all of it has been assembled since around the time I first rode the Spanish bullet train twelve years ago and started mapping a much better world.

I sometimes think that part of the reason why it isn't common knowledge just how much of that world has already been mapped out is because we hear relatively little in the West about what China has been doing on the ground. While much of the world has pecked away at doing less bad, China has charged headlong into the lead in manufacturing the components of a much better world. *Far* into the lead. The pace and scale of the construction of its high-speed rail network—which will reach 70,000 kilometres by 2035 and link every city of more than half a million people in the country—is literally peerless. Add together all the high-speed track everywhere else on the planet, and it equates to barely a third of China's existing network, which is also by far the fastest-growing network.

The numbers for other crucial climate solutions are nearly as staggering. China is the world's top manufacturer and largest installer of wind turbines and solar panels. In 2020 alone, the country added about 50 gigawatts of solar power to its electricity grids, more than existed everywhere on earth just ten years earlier.

China is the world's biggest maker of electric vehicles and the batteries to power them. In five years, just one city, Shenzhen, has rolled 16,000 electric buses onto its streets as part of its public transportation system—about four times more electric buses than there are in any other city in the world beyond China's borders. The superlative stats tally up one after another. Virtually every element of the much better world I've seen in microcosm is now being built in China at industrial scale and breakneck speed. The future I've been chasing is real, and it's mostly made in China.

I don't want to bury this transformation in numbers. They matter—the emissions tally must fall to somewhere very near zero by mid-century, after all—but sums and stats can be numbing, alienating. They are abstractions. A much better world must address people and the kinds of lives they have, the kinds they might yearn for. So let's talk about that future. Let's talk about the value proposition for a good life in the twenty-first century.

"Value proposition" is a business term, and I'll admit it's a little clunky. There are surely more poetic ways to describe the global movement to conquer climate change, more elegant or soaringly spiritual or politically pointed means by which to shorthand it. I could talk about our moral obligation to future generations, the intergenerational debt the world's older generations racked up as they spent lifetimes burning fossil fuels, at first oblivious to the damage and then seemingly powerless to halt it. I could discuss the malfeasance of the fossil fuel industry, its ever-changing strategies of distraction and delay, the heel-dragging of minders of the status quo in boardrooms and legislatures the world over as a catastrophe slowly swept over the earth. I could linger on the injustice of an environmental crisis that weighs disproportionately on impoverished and racialized people in the poorest parts of the world. But in the process of tracking the debate over the past twenty years about how much and how strongly the world should take collective action on climate change, I've come to the conclusion (whether I like it or not) that these ways of talking about solutions reach only limited audiences that are mostly too far from the levers of real power. To truly reshape the world in a generation—as we must—I haven't seen another force as effective as our collective will to improve our daily lives, or at least the prospects of same. So I'll talk instead about a new value proposition.

Here it is: Imagine the suburban ideal, the one that drove so much of commerce and construction in the past seventy years or more. (By one carefully researched estimate, 68 percent of all Canadians and 86 percent of the country's urban residents live in a suburban community.) I don't have to describe it, I'm sure—even if you don't live in a suburban split-level on a single lot with an SUV in the garage, you know how it is. You grew up in one, or your friends did, or your parents live that way now. It's the standard stage set for family sitcoms and cereal commercials, the default setting of daily life.

And this is the very crux of building a much better world—changing the default setting. Not so that, a generation from now, we all live exactly the same lives, but so that the foundations of our ways of life, their most common modes and standard systems, are built from climate solutions. So that daily living is by default emissions-free, or nearly so, and so that our unthinking routines are not slowly making the climate crisis worse every ordinary day.

Very few of us made deliberate choices regarding the fuel in the furnaces that keep our homes warm in winter, or the power source that sends electricity to our air conditioners and refrigerators and TVs and smartphone chargers. When we went shopping for a house or hunting for an apartment, we didn't carefully calculate the most efficient and least polluting routes to our offices and schools and the nearest supermarket. We just looked for the place and amenities that worked best, with the least hassle, that we could afford.

So imagine a new ideal in place of that suburban sitcom set. Call it a reboot, with some of the same trappings as the old twentieth-century classic, but updated, enhanced, thought through a little more carefully. Think of oversized parking lots and decaying strip malls and vacant spaces filled in with denser and more vibrant urban life. Think of a power bill that tracks not just energy consumed but also energy saved and—more often than not, let's say—energy *produced*. Think of a commute free of lurching traffic, a trip to work by train or bike or foot. Think of cars—there will still be many, many cars—with battery packs instead of fuel-burning engines. And think of those engines filling up overnight, when power from sources like the wind is cheapest, and selling power back at a profit while they're parked outside offices all day. And in those offices there will be, in growing numbers, good jobs with bright futures in carbon capture and efficiency retrofits and smart-grid software and who knows what else.

I'm not trying to tempt with a teasing glimpse of a fantasy. I don't believe the revolution—any revolution—is at hand, and I'm not expecting anything like utopia at the end of it all. Human beings will remain the same maddening, amazing, deeply fallible species they've always been, operating in social and economic and political systems that sometimes seem successful only in relation to much worse options. But eras of great improvement do happen. Just a couple of generations ago, much of the world was a frustrating mix of blinding technological wonder and abject horror. Famine, poverty and deadly, incurable disease all cast constant shadows over most people's lives. Wars at continental scale were something like routine occurrences, expected rites of passage. The time interval from then until the present day was not much longer than the interval from now until this value proposition might reasonably be expected to become our new default setting. And the pace of change is already faster.

Daily life is not on the verge of perfection. But it will be *durable* again, insofar as anything wrought by human hands ever can be. It won't be barrelling headlong toward ecological catastrophe. And that is enough, and worth the struggle.

As an Aside: Timelines

In the obsessive climate-policy-wonk circles I've spent the past several years hanging around, there is a reclusive minor rock star by the name of Vaclav Smil. He's a professor emeritus of environmental science at the University of Manitoba, which is a perfect biographical detail for a man to whom I've attached the adjectives *reclusive* and *minor*. Smil writes meticulously detailed books and papers about the global energy system—in the jacket blurb for one of his most recent, Bill Gates claims to wait on Smil's next tome "the way some people wait for the next *Star Wars* movie." (Gates's enthusiasm is the source of a significant measure of Smil's rock star status.) Smil could perhaps best be described as the world's foremost expert on the history of energy transitions. I'll be coming back to his work on several occasions, but for now I want to note what he has to say about the timeline of a global energy transition from fossil fuels to renewable energy.

"The ubiquity and magnitude of our dependence on fossil fuels and need for further increases in global energy use," Smil writes in his 2017 book *Energy and Civilization*, "mean even the most vigorously pursued transition could be accomplished only in the course of several generations." To Smil, not just civilization but human evolution itself is a sequence of increasing dependence on "ever higher energy flows." This dependence is characterized by long periods of relative stasis, followed by intervals of innovation that can be considered rapid only on a geological time scale.

The capture of fire—humanity's original energy transition from pure dependency to control—transpired over the course of a million years or more before becoming widespread about 30,000 years ago. The next great energy transition— "a truly epochal divide in human evolution," Smil writes—was the shift from foraging to farming, which occurred at varying time scales around the world over the course of about 30,000 years and was complete on most of the planet by about 2,000 years ago. Fire remained humanity's main source of non-food energy, and it was fed predominantly by burning wood and biomass such as animal waste and straw, though burning coal for iron making was common in China from the earliest available records. The only reliable source of continuous power in this pre-industrial world was the water wheel, which was as valuable to the ancient Romans (who fed theirs by aqueduct) as it was to the European mercantile economy of the early 1800s.

What came next was without precedent, like capturing fire a million times over. "In 1800," Smil writes,

> the inhabitants of Paris, New York or Tokyo lived in a world whose energetic foundations were no different not only from those of 1700 but also from those of 1300: wood, charcoal, hard labour, and draft animals powered all of those societies. But by 1900 many people in major Western cities lived in societies whose technical parameters were almost entirely different from those that dominated the world in 1800 and that were, in their fundamental features, much closer to our lives in the year 2000.

This, then, is the scope and timeline of the fastest and most dramatic energy transition in human history: the rise of fossil fuels. It transpired mostly over a single century, from the first industrial-scale oil discoveries of the 1860s to the postwar consumer boom of the 1950s. However, the full transition took closer to 200 years, its starting point marked by James Watt's first patent on the steam engine in 1769.

Smil's wary attitude toward the most breathless hype about the pace of the current energy transition is well worth taking into account. That said, the current transition is being conducted deliberately, with considerable international coordination and cooperation, at a time when supply chains effortlessly circle the globe and information blips from one side of the earth to the other at the speed of light. Let's assume for the purposes of this discussion that, while the global energy transition might not be completed in a decade, it is already at the stage of Ford's first Model Ts. The bulk of its transformative work should be possible in an interval similar to the one between those early Ford assembly lines and the birth of the modern suburb in Levittown, New York, in 1947. Fifty years, give or take. Germany passed the world's first legislation designed expressly to catalyze a global energy transition in April 2000. Let's assume, then, that the results of the current transition could feasibly be the global norm by April 2050.

2.0

THE LONG, LOOPING, "LESS BAD" LEARNING CURVE

2.1 A Quarter Century of Reckoning with a 200-Year Mess

Climate cycles are long. There is ice in the Arctic older than the oldest stories in the Bible, and there are deep pockets of ocean water that take thousands of years to fully loop around, carrying carbon dioxide absorbed at the surface to the ocean floor and back again.

Starting in 2009, the scientists who identify, categorize and name geological time spans began to consider whether human activity on earth had left a clear enough footprint to qualify as a new geological interval—specifically, the Anthropocene epoch. Epochs are longer than ages but shorter than periods; they typically last from 5 million to 30 million years. But the Holocene epoch (which began at the end of the last ice age) was possibly brought to an abrupt end after only 11,000 years or so by humanity's rising industrial might.

The geologists examining the Anthropocene decided in 2016 that it did indeed represent a new epoch, although there is not yet agreement on when it began. The dawn of the industrial age is one popular choice, the immutable geologic signature of widespread nuclear testing after the Second World War another. Either way, the massive permanent changes to the earth's natural systems wrought by industry have birthed a new world.

Our collective reckoning with the Anthropocene's signature problem, climate change, has been under way since roughly the late 1980s. The United Nations Framework Convention on Climate Change was established in 1992, and the Kyoto Protocol—the first attempt at coordinated international action—was signed in 1997. So let's say a quarter century in total spent figuring out how big the problem was, how fast it was unfolding, and how exactly we were making it worse. And how to fix it.

At times in that quarter century or so, the progress seemed glacial. Sometimes it even appeared to be moving in reverse. China spent the first decade of the twenty-first century expanding the world's largest arsenal of coal-fired power plants at staggering speed. The United States spent the second decade of the century developing new oil sources at a rate not seen in a hundred years, adding more than 8 million barrels to its daily production by pulverizing shale to make the crude flow.

A quarter century on, it's best to think of the maddeningly uneven, looping work on the problem as not a defeat but a prelude— longer than it needed to be, beset by unnecessary doubts, muddied by self-interest, but a prelude nonetheless. An introduction, a start. The world spent those years getting ready to launch the fastest and most dramatic energy transition it has ever seen. It's easy to look at that as wasted time, but better, I think, to realize it was unlikely that, on their own, a UN declaration in 1992 or a non-binding protocol in 1997 could possibly have triggered the level of political, financial and industrial activity necessary to build a new global energy system. The established order was not going to abandon its comforts readily, and the pursuit of less-bad copies of them was a necessary phase. In retrospect, I think it'll look less like wasted time and more like a short, messy interval on the way to a period of staggering progress.

So let's chart our current coordinates on this optimistic journey, considering what worked and what didn't in those twenty-five unsettled years.

As an Aside: Technical Specs

I don't want to bog down this story too much with numbers and statistics. The engineers and policy wonks love that stuff, they live and breathe it—watts and joules, barrels of oil equivalent (BOE) and tonnes of carbon dioxide equivalent (CO_2e)—but it resonates very poorly outside their circles. Is 100 megawatts a lot of electricity? How does a megatonne of carbon dioxide compare to a gigatonne? How does a megawatt of installed capacity differ from a megawatt-hour of power generated?

I won't be able to avoid energy units entirely, though, so I'll simply vow to do my best to keep the hard-to-contextualize numbers and insider jargon to a minimum. Benchmark it this way if you like: 5 kilowatts is about the size of one of those backup propane generators on two wheels that they sell at hardware stores; one megawatt is about the same amount of power as the diesel engines in two eighteen-wheel freight trucks; and one gigawatt of power is half a Hoover Dam.

I'll also use rate stats here and there, since they are more easily understood. You might not know offhand how much carbon dioxide your house generates each year, but you do know that cutting that number in half is more meaningful than cutting it by a tenth—and that cutting that number to zero is the ultimate goal.

2.2 The View from Svaneke

April 2019. A picnic table with a postcard view of the Baltic Sea. Looking east toward Lithuania, far over the horizon, the afternoon sun setting behind me over the small green Danish island of Bornholm. Left of the picnic table is a smokehouse, a low whitewashed building flanked by five pyramid-shaped chimneys, where herring and mackerel are carefully smoked the same way they have been for more than 150 years. To my right, around a rocky point, a sculptural representation of the prow of a Viking ship stands watch over a small harbour. This is the small town of Svaneke, lauded as the most picturesque fishing village on Bornholm, a quiet, artisanal sort of place that fills with tourists at the height of summer but saunters somnolently through the rest of the year. And I was there, snacking on delectable smoked herring and washing it down with a shot of the island's famed aquavit, because of all that had happened in Bornholm that you couldn't see from Svaneke's harbourfront in the ten years I'd been paying visits to the place and tracking its progress.

There were wind turbines scattered across the island's low green hills and broad farmers' fields. There were district heating plants, powered by burning straw and wood pellets and waste. If you knew where to look (mostly in the larger towns), there were tidy tiled roofs crowned with solar panels, electric vehicle charging stations, appliances connected to the island's power grid by smart plugs that not only delivered power but talked to the grid about current demand and made choices, based on five-minute pricing increments, whether to ask for power right away or wait for a quieter moment and lower prices.

I was in Svaneke that day because the island of Bornholm had spent the preceding decade playing host to some of the most advanced field tests of smart grids and clean energy technology anywhere on earth. And because those tests had by and large succeeded, the Danish government was announcing plans to build two of the biggest wind farms in the world, out in the Baltic Sea off Bornholm's southwest coast, as central pillars in the construction of a fully renewable energy system for the entire country. From island-scale testing to national-scale expansion in a decade. And the herring was still as tasty as ever, and no one sipping a drink on the patio on Svaneke's medieval town square that afternoon would notice a single intrusion on the view, not one shred of evidence of dislocation or calamity.

It's a veritable certainty that Bornholm, along with most (if not all) of Denmark, will be generating all its electricity from emissions-free sources by 2030 at the latest. The entire country will be emissions-free by 2050. And though there will likely be problems caused by more extreme storms rolling in from the sea and more extreme temperatures troubling the Nordic climate, Danes will find themselves on a much more durable footing from which to face those challenges. As a seeker of climate solutions, I'm not sure I've ever surveyed a landscape that could claim to be as near to having the problem solved as Bornholm in April 2019.

I'll return to Bornholm later when I go into greater detail about the twenty-first-century value proposition. For now, though, I want you to feel the Baltic breeze, the contentment that comes from knowing the winds are favourable for the rest of the journey.

I can't remember what exactly I knew about Denmark when I first started looking for climate solutions, but I'm quite sure it was very little. Hans Christian Andersen, stylish furniture, Carlsberg beer. But I googled something like "renewable energy" plus "most ambitious" one day, and that was how I learned that another Danish island, called Samsø, had embarked on an effort to become the country's first "renewable energy island," completely free of fossil fuels. (Bornholm is about 150 kilometres east of the large island of Zealand, where Copenhagen is located, and Samsø is 35 kilometres west of Zealand, between it and the mainland.) At the time I had no systemic sense of how greenhouse gas emissions might be eliminated, no depth of knowledge about how the world's energy systems and energy economy worked. But I had decided to seek out solutions to the climate crisis, and I'd established two rules for myself: to be a climate solution worth talking about, it had to be real, and real people had to be using it. No lab experiments or computer projections, no one-off demonstration projects, no blueprints of perpetual motion machines that would never be built.

In Samsø, the plan was to supply the island's 3,700 residents— all of them—with real emissions-free power. So I went there in the fall of 2005, and I toured wind farms and solar thermal heating plants and got to know a brilliant community-scale energy transition expert named Søren Hermansen. In due course the island of Samsø eliminated all the coal and natural gas from its grid and built enough excess wind power to offset emissions from the oil still needed by the island's cars and ferries. They even built a small sort of school, the Samsø Energy Academy, with Hermansen in charge, because so many energy planners from around the world wanted to see how the Danes had done it. This wasn't that long ago, really. But the idea that renewable power could be the largest energy source on a modern electricity grid and that fossil fuels could be well on their way to retirement— these were fanciful notions on the wild outer edge of possibility back in 2005.

I would come to know Denmark very well in the years after that first visit to Samsø. I spent time on other islands that were steadily reducing their dependence on fossil fuels, gained an intimate knowledge of Copenhagen's unparalleled urban cycling infrastructure, toured industrial facilities where waste products were traded between one factory and another. On one visit I had my family with me, and we went to Legoland. During all that Danish travel, I never once needed a private vehicle—no rental car, no drive to the airport. Which sounds significant to how Denmark became a world leader in the climate-solutions game but is almost incidental. The Danes never stood up as one and announced that to overcome the climate crisis they would have to stop driving cars. The first lesson of Denmark's achievements— the very first lesson I learned on the climate-solutions beat, though it took awhile to sink in—is that the best solutions arise not by stopping what you *don't* want but by seeking what you *do* want. Not by reducing emissions or eliminating fossil fuels but by building a new kind of grid, developing better kinds of transport, assembling a *much better* way of living.

    The main reason Danes are better at finding solutions than the rest of us is that the crisis arrived there a generation sooner, in the form of the OPEC embargo of 1973. At the time, Denmark was one of the most oil-dependent nations on earth, relying on imported oil for more than 80 percent of all its energy use—not just gas for cars but also fuel for power plants and household heating. Gas lines and Sunday vehicle bans became the norm overnight. Some Danes of that generation remember crowding into just a few rooms of their homes during the cold winter because heating costs were too high to keep a whole house warm. And so Denmark started searching for ways to meet its energy needs without importing oil. Along the way, climate change descended as an even graver crisis than the OPEC price shocks. But the Danes did not then decide to unveil the most ambitious climate targets at the United Nations or obsess over their carbon footprint. They asked different questions, the the ones that I suppose are fitting for a nation whose identity is so closely tied to practical, efficient design: What does a better world look like? What tools can be made to move people from place to place and supply energy grids that do the job better than internal combustion engines and power plants that burn oil and coal?

In essence, the Danish solution was to imagine what a world without emissions looked like, and then to pursue that vision relentlessly. A Danish farm machine company called Vestas, for example, became the world's leading manufacturer of wind turbines, because it turned out that wind was the most plentiful renewable energy source in a flat, often cloudy country. In time, as the Danish government's targets for wind power nationwide climbed toward 50 percent, it convened technical experts and academics and big engineering firms like Siemens and IBM to help reinvent its electricity grids. And that is how Bornholm came to be the proving ground for grids that can coordinate the needs of electric cars and smart appliances with the ability of uneven, unreliable breezes to deliver electricity.

The Danes are not yet ready for their victory lap. There are blustery days when all the energy running down Danish power lines comes from the wind, and the carbon footprint of the average Copenhagen resident is about one-fifth the size of that generated by the average resident of Toronto or Chicago. But sometimes the wind doesn't blow, and there are still cars and trucks burning gasoline on every street, and 23 percent of Copenhagen's energy still comes from coal and natural gas. But the pieces of a clean future grid are in place. Denmark ranked first on the global Environmental Performance Index compiled by Columbia and Yale universities in 2020, and it's perpetually among the top half-dozen countries in other global climate action rankings. The remaining problems will be solved.

This is what the Danes have achieved in the fifteen years I've been watching their progress—and what they'll show the world in the next fifteen. And that's why I'm so enamoured of that view from the harbour in Svaneke. You can see a brighter future and a better world so clearly on the horizon.

2.3 On Plausible Optimism

There are only 40,000 people living on Bornholm, not even 6 million in all of Denmark. That's barely a third of Ontario; California alone contains nearly seven Denmarks. Though a single Danish island with no emissions and a clear plan to make it the model for the rest of the country met my initial criteria—*it has to be real, and real people have to be using it*—that wouldn't be sufficient to make a case for a much better world. The world is made of norms and standards and regressions to the mean, not outliers and one-offs and glittering exceptions.

Fortunately, Bornholm is not alone, and neither is Denmark. In fact, nearly all the outliers and glittering exceptions I stumbled upon in my first erratic steps down this path have expanded from their back corners and distant margins toward the mainstream. Renewable energy has been the largest source of new electricity generation on earth since 2015. Wind and solar power, which were among the most expensive energy sources the day I first set foot on a Danish island, are now among the least expensive ways to generate power in most of the world.

The demand for electric vehicles outpaces the ability of automakers to build them, and many of those automakers—including

Volkswagen, BMW, Nissan, Hyundai, Kia and General Motors—have announced plans to make at least as many EVs as any other vehicle type within a decade. Half of Europe, meanwhile, has pledged to ban new gasoline-powered cars from their roads within a decade or two. The hyper-efficient, eco-friendly building designs that once guided construction of the occasional exceptional project—a single office building commissioned by an especially green-minded company, a custom-built home for some deep-pocketed climate-conscious client—are now being woven into the building codes of provinces and states and whole countries.

The ranks of corporations committing to net zero emissions targets are no longer populated primarily by well-known green businesses such as Patagonia and the Body Shop, but also by oil companies (BP, Shell, Total), digital tech titans (Google, Apple, Microsoft, Amazon) and retail giants (IKEA, Walmart, Unilever). Climate politics, not long ago given attention once or twice a year at international summits, has become a daily top-tier agenda item in much of the world. "The world is in the midst of an energy shock that is speeding up the shift to a new order," *The Economist* reported in the fall of 2020. "The 21st-century energy system promises to be better than the oil age—better for human health, more politically stable and less economically volatile."

I think of it now as a plausible optimism. Plausible because there is a certain inevitability generated by such a significant amount of change. A hundred billion dollars per year in Chinese government investments and the sustained efforts of everyone from Walmart to Shell—all of that does not simply dissipate into nothing. Plausible too because there is every reason to expect that the current pace of innovation, implementation and long-term investment (of political capital as well as the financial kind) has guaranteed that the next ten years and the ten after that will bring about much greater change to the climate-solutions realm than the ten years just past. But also plausible—as opposed to inevitable—because those gains remain uncertain. Plausible because it will take more than reasonable arguments and scientific evidence and a pile of favourable data to bring about. Plausible because the waters are treacherous, the sharks and barracudas plentiful, and the navigation through uncharted seas. But there is a route. There is a way ahead. And it leads to steadier ground.

And plausible, finally, because the word lends a tempering tone, colouring the enterprise with a shade of dark uncertainty. Plausible things might not work out as planned. The climate crisis is already a catastrophe, and it's barely begun. The losses are real, and mounting. The world we build with such plausible optimism will not be the one we once had. Tragedy is already as inevitable a part of this plausible future as inspiring change.

2.4 It's Okay to Not Be Okay About It

I think a lot about a man named Charlie Veron, a man I've never met in person. He lives near Townsville in Queensland, Australia. He served for many years as the chief scientist at the Australian Institute of Marine Science, and he is considered the world's foremost coral taxonomist, which means he has identified, categorized and named more species of coral than anyone else. Because scuba diving was invented only eighty years ago and Charlie Veron was the first scuba-diving researcher to study the Great Barrier Reef full-time, it's likely that few (if any) people in the whole history of humanity have spent as much time as he has immersed in the abundant daily life of the world's largest coral reef. He knows parts of it the way other people know their backyard garden.

The reason I think of Charlie Veron so often is because he was the one who taught me how to confront and process my ecological grief. This wasn't the term for it at the time, of course, nor did he impart the lesson intentionally. When I landed in Melbourne for a lecture series in June 2008, there wasn't sufficient awareness about the grave state of the planet's climate for anyone to have yet coined a term for the mix of anger, anxiety, helplessness and despair that many people feel when they look full on at the scope of the crisis. The term began to appear in scientific journals around 2017 or so—one researcher began to codify its impact on the Inuit as they witnessed radical change in their Arctic homeland—and not long after that, the mainstream press picked up on it as a sort of lifestyle trend. "It's not just anxiety that shows up when we're waking up to the climate crisis," Britt Wray, an expert in science communications working on a book about ecological grief, told the *New York Times*. "It's dread, it's grief, it's fear."

I'm not even sure where my own ecological grief came from. I don't recall some indelible moment when I realized the planet— the one I would hand to my children—would bear permanent scars of human recklessness and might have a climate so volatile that it would pose an existential threat to civilization. I worked for Greenpeace in the summer of 1993, but we mostly talked about the clear-cutting of old-growth forests. I took an undergrad course in environmental history the following year—it was such a new and marginal topic that the course wasn't yet offered every year. Around the same time I started reading some of the first mainstream books and articles on global warming, and later, when I became a journalist myself, I was working at *Time* magazine—writing about high-tech digital wonders—when it ran its first polar-bear-in-peril cover in the fall of 2000. (The bear seemed to be stranded on a pillar of ice cleaved from the frozen land mass behind it. The headline read: "Arctic Meltdown: This polar bear's in danger, and so are you. Here's how global warming is already threatening the planet.") I wrote a short profile of a climate scientist named Andrew Weaver, and after I'd turned off my recorder there was something in the hushed, anxious way he talked about the implications of his work that reminded me I should be paying closer attention, that the situation was growing more dire by the day, that climate change would define the twenty-first century, and that I should be using whatever resources I had at my disposal to bring more attention to the emerging crisis.

Sure, there was all that. But I don't remember anything like the full weight of despair. My grief came to me in quiet increments, a sort of clouded spectre at the edge of my thoughts, a motivating force too horrifying to fully comprehend. But then I landed in Melbourne and there happened to be a story in the newspaper about Charlie Veron's new book, a natural history of the Great Barrier Reef that was the culmination of his life's work. Veron had been too focused on the teeming life under the ocean's surface to spend much time on the crisis in the atmosphere, but in the writing of the book—so the story in the Melbourne paper explained— he'd discovered that climate change posed an existential threat to the Great Barrier Reef itself. To every reef on earth, in fact.

Here's how *The Age* summarized Charlie's concern that day, June 7, 2008:

Twin assailants, both creatures of climate change, threaten the reef and oceans more generally. The lesser of these is the warming of the water, which turns the single-celled algae on which corals rely for their sustenance toxic, compelling the coral to expel them and probably die—the event known as coral bleaching—or to keep them and certainly die.

The worst bleaching events of history will become commonplace by 2030, says Dr. Veron, and by 2050, "the only corals left alive will be those in refuges on deep outer slopes of reefs. The rest will be unrecognisable—a bacterial slime, devoid of life."

The even greater threat is ocean acidification—the dissolving of carbon dioxide into the sea, forming weak carbonic acid. This is the climate change frontier to which science is swinging increasing focus, as alarm grows at the threat it poses to marine ecosystems and to human food supplies and economies.

I was sitting jet-lagged in the sun on a patio on the Melbourne harbourfront when I first read that. I had no idea how to process it. And I wasn't used to getting stuck on where to file new information. Synthesizing news items was a significant part of my profession, but I was struck numb by this one. Which is, as I came to understand only much later, one of those things grief often does to people.

It seemed necessary to take some sort of action, and as luck would have it, I was in Australia on a special airline pass that allowed me several internal flights. So I flew to Brisbane and then up the Queensland coast, to an island chain studded with beach resorts called the Whitsundays. I took a catamaran to a spot called Hardy Reef, one of about 3,000 individual reefs in a staggeringly scaled symbiotic chain that stretches more than 2,000 kilometres up the Australian coast and beyond. A floating platform was moored on Hardy Reef for tourists, and I went diving. I told myself that I wanted to see the Great Barrier Reef in all its impossible wonder, this largest living system on earth, this teeming belt of aquatic life twice the size of Florida, in order to know how to describe its value, its magnificence. In order to help in some small way to save it. But there was that grey cloud at the edge of my thoughts, the one that said the real reason I'd come to the reef was so I could see it before it died. So I could, at least, help properly identify the victim.

The April 1, 2021, edition of the *Sydney Morning Herald* provides a grim update. "Barrier Reef Now 'All But Doomed,'" reads the headline. The story discusses a report from the Australian Academy of Science, indicating that at least 70 percent of the Great Barrier Reef "would cease to exist as we know it" as a result of 1.5°C global warming, according to one of its authors. At 2°C, as little as one percent of the reef would be expected to survive.

In 2016 the Great Barrier Reef endured the most intense bleaching event in recorded history: more than 90 percent of the immense system's 3,800 constituent reefs were damaged. Then in 2017, before its delicate corals had even begun to recover, another ferocious bleaching event occurred. Coral bleaching was unknown before 1980, before the oceans warmed. Back-to-back events on such a catastrophic scale were far beyond the Great Barrier Reef's ability to cope. The damage is permanent, and deepening.

There is nothing to be done, really, about a sense of loss on that scale. It's okay to not be okay about it, to turn it over and shake it around and never be completely at peace with it. How do you mourn the passing of something hundreds of times older than you, a whole ecosystem the size of a European country, whose dying—if that's even the right word—will surely stretch beyond your lifetime? How do you calculate the loss? Count up this shoal of fish?

Those anemones? The prongs on that staghorn coral formation? It's impossible to fathom in full, just as climate change is impossible to fathom in full.

My first thought was that this "ocean acidification" phenomenon, though largely unknown, was so catastrophic in scale that surely its story should be told. I read Charlie Veron's book, *A Reef in Time*, and spoke to him by phone a couple of times. I went to Florida for a conference in the spring of 2012, and I took a couple of extra days to interview scientists working on ocean acidification at the National Oceanic and Atmospheric Administration's laboratory near Miami. What struck me was the mix of excitement and dread those researchers were feeling. They knew better than anyone that climate change was altering the world's oceans forever, but the acidification problem also created a rare opportunity for scientists working in the twenty-first century to conduct basic field research at the level of gathering baseline pH readings of ocean water and tracking their fluctuations.

It had been known for decades that the oceans absorb carbon dioxide from the atmosphere—roughly a quarter of all the CO_2 generated by human activity—but the mechanisms by which it was making the seas more acidic and the nature of the deep-ocean carbon cycles involved were virtually unknown. The term itself— *ocean acidification*—had only been established and clearly defined in a 1999 research paper. Marine scientists retell in hushed tones the tale of how that paper's lead author, Joanie Kleypas, had been attending a climate change conference the previous year when she came to the shocking realization that her research revealed impending catastrophe for the world's oceans. She was so shaken she fled to a washroom to be sick. The science was that new, that raw, that deeply disturbing.

I finagled a fellowship in the fall of 2012 to attend a major international conference on ocean acidification in Monterey, California. The science was mostly way over my head—careful technical presentations, one after the other, on aragonite saturation of coral reefs and the ocean's biogeochemical feedback loops. One afternoon there was a networking session with cocktails. It was a typical conference schmoozefest, though I felt slightly more at home in that awkward, geeky crowd than I usually do among the polished networkers at corporate events. I fell into conversation with one researcher about how soon the acidity of the oceans would pass the critical threshold beyond which the algae living on corals could no longer fix carbon dioxide into limestone reefs—the point of no return Charlie Veron had warned about in that Melbourne newspaper. There was much debate in scientific circles about exact dates. Was it, as Veron asserted, merely a few decades off, or would it take a century or more yet? And would the trigger point be the same for every reef? (Likely not. Reefs in different parts of the world have evolved levels of resilience and adaptation as varied as the people who inhabit different climates.) This scientist told me that, as far as the climate crisis was concerned, ocean acidification was a second-tier problem. By the time we reached the level of acidification required to instigate this catastrophic dying, he assured me, life on land would already be such chaos that the oceans would be the least of our concerns. Such were the cold comforts of the climate science beat.

I spent several years after that conference trying to turn this dark knot of knowledge and anxiety I'd amassed around ocean acidification into a coherent story. I thought I could build the narrative around Veron, his own shock of awareness and mounting grief, his desperate attempts at activism. There is a photo, widely distributed, of Veron floating above a patch of lush Great Barrier

Reef coral, his sharp eyes hidden behind a dive mask, unfurling a banner reading "KEEP THE REEF GREAT" over the Greenpeace logo. It's the kind of stunt environmentalists use to raise awareness about the clear-cutting of old-growth forest or the strip-mining of a wilderness. But there was nothing much anyone could do with a single protest campaign—no matter how large, no matter how righteous the message or how much support it generated—to stop the steady flow of industrially generated carbon dioxide into the sea in a timely fashion. This was the futility of the climate crisis, the core of climate grief. Charlie was right, and he was doing the right thing, but it didn't matter much at all.

In the years since I first learned of his work in that Melbourne newspaper, Charlie wrote a beautiful, exuberant, ultimately haunting memoir called *A Life Underwater*. His grief for the decline of the coral reefs he'd spent his life studying permeates every page, intertwined with his love for his work and the tragic loss of a young daughter to a drowning accident. "I had imagined I would be gone before the worst of my fears were put to the test," he writes in the book's conclusion, "but now I'm afraid this may not be so."

I think it was while reading Veron's memoir that I clearly understood that the story of the peril of the oceans is not mine to tell. I haven't spent hours and hours in the depths communing with the reef. I visited only once. My real communion on this journey has been with the tinkerers and technocrats who have been urgently scrambling to build a better world, one that just might emerge fast enough to save some reefs, if not all of them. Immersing myself in that work has been my way through climate grief to a plausible optimism. So that's the story I'm telling.

2.5 The UN Won't Make It Okay (and That's Okay)

I wish the science mattered more. There's an ache in my gut, like something poorly digested—a sort of subgenre of my own climate grief, maybe—born of a clear lesson I've learned while covering the climate beat for nearly twenty years. People, at least in groups, do not make major decisions about how to live their lives based on a solid understanding of scientific facts well articulated by experts. (This point was harder to make before COVID-19 and its brigades of opponents to masking and life-saving vaccines emerged in the daily news for months on end.) The way people make big collective decisions, for want of a better mechanism, is through politics. And that is not a realm guided by science. It's *informed* by science (sometimes), but not guided by it—ever. There's no need to qualify that, to shade it with greater nuance. That is simply not how it works. I'm sorry that's the case, I truly am.

I've now witnessed at least four waves of mounting, cresting and dissipating hope in the international political process that began with the first Earth Summit in Rio in 1992. Each time there is gathering excitement, a sense that *this* time, with the weight of the latest information or the volume of the street protests or the ferociousness of the most recent natural disaster, there will be dramatic action taken at a UN conference—or in its wake, at least. Something, anyway, commensurate to the scale of the anxiety and grief.

There is an obvious logic to shifting the urgent despair arising from a global crisis to the nearest thing we have to a global solutions agency. The United Nations has convened many important debates on the biggest geopolitical issues of the day. It's the forum the world's nations have used to develop rules and consequences for war crimes, to address global poverty, to protect the most valuable pieces of humanity's collective cultural heritage, to set goals for twenty-first-century development and measure progress in their pursuit. It has been engaged in the climate crisis since Rio and has outlined some of the world's most ambitious goals for addressing it.

The Kyoto Protocol, introduced in 1997, was the first major international agreement on how to cut greenhouse gas emissions. The UN climate summit in Copenhagen in 2009 was expected to address the failing Kyoto process, but instead it ended in a maddening gridlock. The Paris climate talks in 2015 produced the nearest thing to universal agreement on the gravity of the crisis that the world has ever seen. And the scientific declaration from the UN's top climate science organization (the Intergovernmental Panel on Climate Change, or IPCC) in the wake of Paris defined in the starkest of terms the importance of limiting global warming to 1.5°C. That triggered a global wave of renewed protest, led symbolically by a Swedish teenager whose blunt, angry rhetoric cut to the core of the crisis and demanded that the world's political leaders "follow the science." As Greta Thunberg's fame mounted, she embarked on a solemn quest to spread her message around the world, culminating in a speech at the UN in 2019. "For more than thirty years," she told the assembly, "the science has been crystal clear. How dare you continue to look away and come here saying that you're doing enough, when the politics and solutions needed are still nowhere in sight."

If the UN is our largest political body, at least in terms of geographic range, then it perhaps makes intuitive sense that it would formulate the response at a global scale. But politics, alas, is a great many things—emotional, reactionary, self-serving, transactional— before it is logical. This is, as the saying goes, a feature, not a bug. Political institutions don't move rapidly in response to the latest data on any subject, particularly one as vital and lucrative as energy production, because they were never built to do so. At their very best (and political institutions are only occasionally at their very best), democratic political institutions are designed to find consensus among competing interests, not to execute the recommendations of scientific reports.

At Paris in 2015, the UN was as near to its best on the climate crisis as it has yet managed. The pledges of the 194 signatories to the Paris agreement came closer to the scale of the problem than any previous UN summit. But goals are not programs, and the strongest arguments don't inevitably win election victories. And so, in real terms, the UN process has taken a quarter century to bring the world together in a voluntary non-binding agreement with no effective mechanism of coercion and no incentives for compliance. There is only a unified statement of best intentions with a smattering of governance piled around it, in the hope that it will be used to create at least some change. If the UN had any real authority to compel action, it would have tried to exercise it by now.

I don't mean to suggest that the Paris climate agreement doesn't matter. It was a historic landmark in the pursuit of climate solutions, a clear and unanimous declaration from virtually the entire world's political leadership that the energy transition under way was inevitable, that the twenty-first century's energy system would be substantially different from the one that held sway over the twentieth. But the UN process is not driving that transformation. It's not setting the pace or even articulating the goals most clearly. And it isn't doing so because it can't do so, for two connected reasons.

The first is that the UN is not a binding world government. It has never had the authority to oblige a single legislature to do anything at the level of real policy. That's by design, and it's a good thing. The world would emphatically not benefit from a binding world government. The complexity and scale of such a body would render it totally incapable of responding to the real needs and desires of people at a community or even national scale. When climate activists arrive at an event like the Paris climate talks and demand stronger targets and an agreement with real power, they are asking an invertebrate to grow a backbone. It's not that kind of animal.

The UN can give voice to our greatest aspirations and enable sanctions for our worst sins, but it can't rewrite 194 energy policies or force regulations upon the whole world through 194 environment ministries. The IPCC—the logical arm of the UN climate process, if you will—is a working group of climate scientists. They examine the vast and complex research and data detailing the state of the world's climate, and they produce reports outlining the status of the current crisis and the range of potential future impacts. It's invaluable, heroic work. But they are not policymaker or energy experts. They can't manufacture more and cheaper solar panels or develop public funding models for investment in emissions-free industries. That's not their focus and it's far outside their areas of expertise. And thinking they or the UN climate process in general could do so represents a fundamental misunderstanding of the nature of the United Nations, the politics of climate change and the most basic fact of the crisis itself. Which is that climate change is not an environmental issue.

I'll repeat that: *Climate change is not an environmental issue.* This is the second reason why the UN can't drive the energy transition. The climate crisis *contains* a number of massive, global-scale environmental issues, including global warming, extreme weather, ocean acidification, species collapse, biodiversity loss, drought and desertification. But it is not, at its core, an environmental issue. It is, rather, the biggest, baddest collective action problem humanity's ever faced, which by definition means it's a crisis not of environmental processes but of human ones. And solving it is fundamentally not about agreeing to reduce emissions but about providing irresistible incentives to accelerate the global energy transition. These might sound like two sides of the same coin, but they are not. They are as different as modelling weather patterns and setting up a factory to produce photovoltaic cells at industrial scale. This misunderstanding is one of the primary reasons why climate politics and climate action in the years of the less bad prelude have been so prone to frustration and despair.

One of the sharpest observers of climate politics throughout these drawn-out prelude years has been an American writer named David Roberts, who started at Grist (one of the internet's first dedicated climate news sites) and moved on to Vox, the "explainer journalism" pioneer. He's probably been my most reliable guide in this muddy, churning swamp of advocacy, activism, politics and propaganda. I wonder if that's because he didn't come to the beat as an environmental activist or an in-the-loop political journalist. What I mean is he didn't arrive at the topic of climate change with an established narrative to fit it into. And because of that, he's been particularly skilful at seeing how the climate narrative got subsumed into old political battles and grudges.

"I'm not an environmentalist and these aren't environmental challenges," Roberts wrote about climate change back in 2010. "The solutions that American environmental politics are capable of producing are not commensurate with the scale and scope of the challenge climate change represents. A clear understanding of that challenge renders comically absurd the notion that it can or should be the province of a niche progressive interest group. It's just too big for that."

Roberts's concern then was that by addressing climate change through an environmentalist lens, climate advocates would remain trapped in the constrained world of the "movement politics" launched in the 1960s. Climate change would be seen as a single narrow problem at the margins of mainstream interest, a boutique issue of serious concern only to the usual suspects on the protest-politics left, to be addressed (if at all) well after heavyweight topics like the economy and foreign affairs. What's more, this long-established narrative is a rigid frame that positions environmental concerns, no matter how grave, as opponents of economic health and individual livelihoods.

Thinking of climate change as primarily an environmental problem is not only imprecise—it is primarily a problem of how human populations make and use energy and how they are organized economically and industrially—it's also a political trap. Conservative opponents of strong climate action and incumbent fossil fuel producers alike have long welcomed and encouraged this framing of climate change as a contest between environmental and economic health. Why? Because it means they can continue to balance their (essential, higher-priority) work of producing energy against the environmental damage it might be causing, reducing the catastrophic changes their products are making to the basic composition of the entire earth's atmosphere to a "special interest" issue—in the same category, regardless of magnitude, as concern for the spotted owl or the contents of a household garbage bag bound for the local landfill.

Another of the wisest people I've encountered on the climate-solutions beat is a British climate campaigner and writer named George Marshall, whose 2014 book *Don't Even Think About It* is a vital primer on the psychology driving people to disengage from concern about climate change. I first met Marshall at a symposium for journalists in Germany back in 2005, and he was already obsessed with what he was calling the "psychology of denial"—the way most of the general public failed to feel sufficient urgency and anxiety about climate change. A significant part of his explanation for that phenomenon has to do with the same environmentalist trap that worried David Roberts. Marshall has spent most of his working life immersed in the green activist world, and he's blunt in his assessment of the movement's limitations. Environmentalism, he writes in *Don't Even Think About It*, provides "no community of belief" and "no social mechanism for sharing it. . . . If climate change really were a religion, it would be a wretched one, offering guilt and blame and fear but with no recourse to salvation or forgiveness."

Take that and compare it to Roberts's call for the climate crisis to "transcend the environmental movement—and movement politics, as handed down from the '60s, generally. . . . It needs to become a shared concern of every American citizen regardless of ideological orientation or level of political engagement. That is the only way we can ever hope to bring about the urgent necessary changes."

And how has that movement responded to its years of continued frustration on the margins? Alas, too often with yet more guilt and blame and fear, alongside a sporadic pursuit of just the right catalytic event or rallying cry to overcome its own limitations. A sort of magical thinking, in other words.

2.6 War Footing and Magical Thinking

For nearly as long as we've been grappling with the climate crisis at an international scale, there has been a parallel hunt for the right metaphor to describe the scope and pace of change that meeting the challenge requires. At times this belief in the power of finding the exact set of words to convey the enormity and urgency of the task has seemed almost messianic, as if just the right terminology could, all on its own, make the solutions manifest. But much as goals are not programs for reaching them, slogans and banners are not the means to make the transformations they promise happen. SAVE THE PLANET! STOP GLOBAL WARMING! KEEP IT IN THE GROUND! CLIMATE ACTION NOW! Great. *How?*

In my years on the solutions beat, I've seen climate change likened to the civil rights movement in the United States, even to the global crusade to end slavery in the eighteenth and nineteenth centuries. Man-on-the-moon metaphors often recur, from the Apollo Alliance, launched in 2003 by American unions, politicians and environmental groups, to the Global Apollo Programme, a 2015 proposal by British academics and business leaders to make solar power cheaper than coal in ten years. (Solar power is cheaper than coal now in a rapidly expanding swath of the world, though few would credit Britain's Apollo Programme for the achievement.) For a time about a decade back, the prime minister of Norway took to referring to a project that was attempting to rapidly commercialize the process of capturing carbon dioxide emissions at natural gas power plants as "our moon landing." (Norway missed the moon; the project has been stagnant for years.) Progressive American lawmakers, meanwhile, have more recently proposed a Green New Deal, likening climate action to escaping the Great Depression.

No other climate action metaphor has proven as compelling, though, as the "war footing." If the task calls for a global-scale, all-in, maximum-speed movement to shrink greenhouse gas emissions to zero within a few delirious years, then what if the world's great powers were to switch over their entire industrial might to solving the climate crisis as quickly as Allied economies reoriented themselves to fight the Second World War? Imagine factories retooled for green power and clean technology, a global army of citizens deployed to install solar panels and erect wind turbines, vast pools of capital poured into energy storage facilities and electric-vehicle charging stations, commuter trains and bicycle lanes and hyper-efficient everything. If we have the tools, then why pursue anything less than total war?

I can certainly understand the appeal of the metaphor. Resolute action on climate change at such a scale would be enticing, intoxicating, sweeping away the shell shock of climate grief with a Normandy invasion and the promise of a liberation march. Imagine gathering children and grandchildren at our feet, perhaps in the climate-calmed world of 2040 or 2050, to recount stories of the elite international corps of engineers and technicians and the home-front factory workers and tenders of victory gardens who beat back the tyranny of greenhouse gas emissions and built a new postwar economy as vibrant as any the world's ever seen. Who wouldn't love to be part of a story that tidy and uplifting?

When I think too long in this vein, I find myself pondering what political mechanism could trigger it all. This is where, for proponents of the war footing approach, magical thinking really begins to seep in. To call for a war footing is, at its root, to dream of perfect political efficiency and maximum control. It is to conjure up a single lever of power obliging a coordinated, sustained global-scale effort to address a crisis that is palpable to most people only in sporadic, geographically discrete ways—a crisis that has been deeply divisive for the politics of nearly every country that has begun to take on the task. A war footing is an erasure of the need to win support, build consensus, argue and compromise. (It's an erasure, as well, of the real history of the Second World War, during which consensus was far less universal than it has been portrayed in retrospect, and corruption, coercion, error and pure greed were far more prevalent.)

More than anything, though, the war footing metaphor is a symptom of the climate movement's roots in environmentalism, which has often distrusted, if not outright shunned, the crass short-term thinking and craven vested interests of politics. Thoreau quit the daily grind of society. Sierra Club founder John Muir was much closer to a preacher than a politician. Greenpeace skipped past the political arena to direct action and media manipulation. A war footing promises the same fluid dodge past the morass of politics, straight into a climate-driven global energy transition. No ill-informed voters, no negotiation or compromise. Just action, pure action, as simple and direct and irrefutable as the storming of a beach on the coast of Nazi-occupied France.

A grassroots activist organization called The Climate Mobilization, which organized protests outside the UN headquarters in New York as the Paris accord was being signed in early 2016, has demanded that the American government "launch a WWII-scale effort to transform the U.S. economy on an emergency basis." In Britain three years later, a prominent policy think tank created the "Environmental Justice Commission" to push for stronger climate action. It was headed by former Labour Party leader Ed Miliband, who told reporters at its launch, "Politics needs to be on a war footing to deal with this enemy." Similar rallying cries for wartime economic command and control, home-front mobilization and Dunkirk-spirited society-wide action have come in a steady drumbeat from activists, academics and politicians around the world.

From whence, though, would a climate war declaration even come? The UN has already demonstrated its inability in that regard. Which single nation's government—itself presumably somehow unanimously convinced of the necessity of such sweeping action, and unwavering in its commitment for years on end—could compel all the others? Or any others, really? By what means? Who would issue orders, who would draw up the plans, and who would execute them? And what of the ineptitude, cowardice and political self-interest that have slowed the process and earned the world's would-be leaders such ire in the most committed climate action circles? Who would sweep that all aside? *How?*

I'm not asking these rhetorical questions for sport here. I wait on the answers every time the topic recurs. In February 2020 (just before the world learned in unequivocal terms how quickly nature can compel change on a global scale), I attended a conference in Vancouver called Globe 2020. For climate and energy wonks in Canada, the Globe conference is one of the biggest dates on the calendar—the kind of event where the federal government sends cabinet ministers to make major climate policy announcements, oil company CEOs sit on panels with climate activists, and electric-vehicle makers display their latest and greatest cars.

At Globe 2020, one of the keynote speakers on the first morning was David Wallace-Wells, an American journalist who'd written the latest bestseller on the imminent climate-caused peril of humanity, a book called *The Uninhabitable Earth*. I think this was roughly the fifth recurrence of the Armageddon narrative I'd encountered in my twenty years on the climate beat, following Al Gore's *Inconvenient Truth* documentary in 2006, books by Tim Flannery (*The Weather Makers*) and Jared Diamond (*Collapse*) in roughly 2007 to 2009 (they became prominent in different circles at different times), Naomi Klein's *This Changes Everything* in 2014, and the wave of protest and anguished press coverage in the wake of the IPCC's special report on the climate impacts of warming beyond 1.5°C in 2018. (Greta Thunberg's first humble solo *skolstrejk för klimatet*—"school strike for climate"—happened that August.) There are surely others, and I don't mean to dismiss the collective story they tell. The work of activists and anguished authors has been vital in pushing the climate crisis up the agendas of decision-makers around the world. But there's always one answer I find myself waiting on each time—the one that explains *how*.

Wallace-Wells was a composed and passionate speaker at Globe 2020, his speech short and punchy like a TED talk. He began with a careful recitation of the many inconvenient truths—the climate crisis already upon us, catastrophic disaster under way and worsening, a humanitarian tragedy of flooded cities and climate refugees all but guaranteed in our far-too-near future. But the problem, he argued, was ours to solve. "The main driver of climate change is human action," Wallace-Wells said, "which is to say how much carbon we put into the atmosphere. And our hands are collectively on those levers. Which means we can write a different story if we choose to. And not just can—*will, must*." Then he chastised Canada's prime minister for continuing to support oil pipelines and suggested we were all "living in denial" about the climate crisis, locked in selfish, nationalistic, hypocritical political structures that only intensified the problem.

"For far too long," Wallace-Wells said, "we've defined our goals in politics through what we considered politically possible. Which means we were always working within paradigms that were established in the past under different conditions rather than building our objectives out of what we knew morally and scientifically was necessary. We can't continue that way. We need to change that paradigm."

And then he was done, and I sat in the dark, echoing conference hall and asked myself again: *How?* As ever, the question was essentially unanswered.

It's not that I disagree with the intent—of Wallace-Wells, of any of the alarm-calling writers and speakers who've long kept this apocalyptic story at or near the centre of the climate crisis discussion. It's the wishful thinking that I get stuck on. Wallace-Wells pointed to the Sunrise Movement and the recent surge in youth engagement and protest it emerged from as the source of his optimism, and I agree that the waves of activism by young people around the world who are fighting for their future have generated the most invigorating force in climate politics in many years. But youth activists can't fully answer *how* any better than Al Gore or the IPCC's scientists could. In the specific case of the Sunrise Movement, it has helped move the climate crisis very quickly to the top of the agenda for the Democratic Party in Wallace-Wells's native United States, which is finally attempting, under President Joe Biden, to enact an ambitious federal climate plan. But that plan has already begun its descent into the churning morass of a House of Representatives controlled by much more moderate Democrats and a Senate partially handcuffed by straight-up climate-denying Republicans. It might change the paradigm—a little. But it won't embed what is "morally and scientifically necessary" into the bedrock of American government.

I've come to think of the beginnings of these statements as a sort of tell. *We need, we must, we have to, we should, we could.* These are ways to indicate what their author would like to see happen, not ways to explain *how* to make it happen. This recurs time and again in these discussions, in the calls for war footings and paradigm shifts and strict adherence to what the science tells us. Wallace-Wells dreams of a new politics written by the Sunrise Movement. Greta Thunberg hopes that the clear, logical certainty of her arguments will compel the world's political leaders to act in ways they never have before and direct their institutions to do things they've never done. Al Gore reckoned that if enough people saw compelling data in just the right format, the debate would end in victory for the forces of science and logic. And so it has gone for well over a decade.

I see the same wish fulfilment in the slogans and memes that have ignited the climate-advocacy community of late and then fizzled in the swirling winds of the political arena. TWELVE YEARS LEFT TO SAVE THE PLANET. JUST 100 COMPANIES RESPONSIBLE FOR THE MAJORITY OF EMISSIONS. STOP FOSSIL FUELS. One after another, these reduce a staggeringly complex scientific, economic and political problem to what might sound like a single decisive step. It's a compelling rhetorical trick. It's also fiction.

2.7 The View from Berlin

In April 2019 I attended Berlin Energy Week, an expansive show-case for Germany's leadership in the pursuit of climate solutions, hosted by the German Foreign Office. The event's centrepiece is an international conference called the Berlin Energy Transition Dialogue (BETD), and the host city is an obvious fit for such a discussion. My pursuit of climate solutions had led back to Berlin again and again. I've come to think of the city as the epicentre of the climate-solutions world, its primary axis, a place where I've learned many crucial lessons about what a shift of such magnitude looks like at street level.

In Germany, they call this epochal shift *die Energiewende*, which translates as "energy transition," a term that has only recently come into common usage outside Germany, even among climate policy wonks. I first heard it in 2008 from an executive in the industrial heartland of Germany's solar industry, south of Berlin. The country's improbable solar boom was at its peak—a wave of frenetic industrial activity triggered by Germany's ambitious renewable-energy legislation, which almost single-handedly rationalized the entire solar business. It caused such feverish investment that the TecDAX index, which tracks German technology stocks, was being referred to at the time as the "solar DAX."

The solar boom was part of a plan to eventually eliminate all fossil fuels from the German electricity grid, the executive told me—adding, with a sort of mildly embarrassed shrug, "This *Energiewende*, we call it." There was ironic distance in his intonation. Maybe it was empty buzzspeak, this term. Maybe it was some German quirk. It translates quite readily to English, after all: "energy transition." Why be so precious about using the German version? A decade later at BETD, the German government had turned the phrase into a logo on the welcome banners for its marquee energy conference. "ENERGIEWENDE—SWITCH TO THE FUTURE," they read, with a stylized German flag waving between the two halves of the compound word.

The conference was a concerted effort by several German government departments—chief among them the Foreign Office and the Ministry of Economics and Energy—to showcase their progress on the *Energiewende* and encourage similar efforts around the world. It's telling that the German government, which has had this transition as a top priority as long as any national government on earth, does its cheerleading through the energy and foreign affairs ministries. This isn't an environmental issue in Germany, at least not primarily. It's a central question of economic development and international trade.

The venue underscored the symbolism, taking over the entire ground floor of the stern old German Foreign Office building and adjacent modernist extension. The complex's main building was the Reichsbank during the Nazi era, the site from which the Germans financed their war; the multilevel basement was once filled with vaults full of stolen gold. Later it was the headquarters of the Central Committee of the Soviet-era Socialist Union Party, the uncontested rulers of communist East Germany. After reunification a new section was added, an extension known for its airy atria and zealous energy efficiency—hallmarks of a new, more open and less destructive era for the German state.

The BETD's main plenaries transpired in the grand tellers' hall of the old bank building—a significant upgrade in gravitas from your typical Sheraton conference centre. And the event attracted a much weightier crowd than the norm, as well. There were green entrepreneurs and veteran climate advocates, of course, but BETD also brought in senior energy and climate policy officials from around the world, including at least fifty cabinet-level ministers and secretaries, as the German Foreign Office was quick to brag. This was the vanguard of the global energy transition, at least in an institutional sense, and as near to a gathering of the would-be planners of a worldwide shift to a war footing as I've encountered to date.

"We now see a global energy transition," an official with Germany's economics and energy ministry announced. "And we are very happy to see that, because we are convinced that the energy transition is in fact the future."

The spectre of the climate crisis loomed over the proceedings, tempering high-minded talk of carbon-free grids by 2030 and next-generation biofuels with much less rousing assessments of our progress to date. Fatih Birol, executive director of the International Energy Agency (IEA) and a strong advocate for action on climate change, delivered what he called "an X-ray of the global energy situation." His diagnosis was grim. Birol explained that 2018 had been "a remarkable year." Renewable energy was expanding as never before and getting cheaper by the day, the largest source of new power on the world's grids. Euphoric news, to be sure. But still—darkness encroaching—the renewable energy business wasn't moving fast enough to keep pace with greenhouse gas emissions, which had reached a new peak in 2018, driven by the largest spike in worldwide energy demand in a decade. "There is a growing disconnect," Birol said, "between political statements, targets, and what is happening in real life."

There had been some great achievements—the rise of cheap, ubiquitous renewable energy was a true marvel—but a long, perilous march lay ahead. Transportation? Emissions from heavy industry? The whole world of steel and glass and concrete that is our built environment? All have barely begun to be addressed in any kind of comprehensive way.

To watch Birol speak under a banner reading *"Energiewende"* was to gaze upon a glitzy snapshot of dark euphoria. On the one hand, I found myself marvelling at the sheer grandiosity of the event. Little more than a decade earlier, my reporting on the energy transition, even in Germany, involved slogging through the back hallways of obscure government sub-departments. A sort of skeptical cloud loomed. The Germans had stampeded toward solar panels and wind turbines far too eagerly, even recklessly—this was the common wisdom beyond its borders. Their policy tricks and industrial subsidies would surely lead to a gilded boondoggle, and maybe take the whole German economy with it. In 2006 an analyst at the Rocky Mountain Institute—a solar expert, no less—informed me scoffingly that at Germany's inflated rates he could make a profit by putting monkeys in cages and getting them to run on treadmills to generate power. Even at the edges of my conversations with Germany's own *Energiewende* planners, I sensed hesitation. Something like this had never been tried. Who knew if it could really work?

The morning that the IEA's executive director took the stage in April 2019, more than 40 percent of Germany's grid capacity was coming from renewables, and Angela Merkel had earned the nickname "climate chancellor" for her forceful lobbying of reluctant EU member states for stronger climate action. And here was Birol, head of an organization created initially as an adjunct of the oil industry, onstage among many other dignitaries in a grand hall at the heart of Europe's most powerful economy, his message echoing activist calls for much stronger action. Chastising the many politicians in the crowd for falling short, singing the praises of solar panels and wind turbines. The head of the damn IEA demanding more climate action now? That alone was euphoric in its way.

There was no denying the dark aspect, though. I'd spent the past decade watching as one forecast or prediction of the potential for clean energy growth after another—many from the IEA itself—was not just surpassed but thoroughly trounced. One European nation after another was expanding its solar and wind capacities with dizzying speed. China was conquering the entire sector with a subsidizing zeal so intense that a German solar executive once told me you couldn't buy ingots of the raw silicon used to make solar panels for the price at which the Chinese were selling finished photovoltaic solar cells. Electric car sales booming, high-speed track laid across half of Asia, green building standards fast becoming the norm—all that, and yet Birol's numbers were unequivocal. Emissions were still rising. The world had not even begun to turn the corner on the crisis, much less solve it.

This is where support for the war-footing approach springs from. Looking outward from the maelstrom of the climate crisis— the rapidly melting Arctic, the deadly wildfires and hurricanes, the floods and heatwaves—whatever process had brought us to this horrific reckoning must have failed us, as surely as appeasement and isolation failed to stop the spread of fascism in the 1930s. What was left but to declare total war, to fight in the fields and in the streets and never surrender?

I carried the prospect of a war footing with me throughout the six days I was immersed in high-minded energy transition talk in Berlin. I tested it mentally against the conversations that swirled around me, from conference hall to evening gala to clean energy demonstration site, swerving from breathless enthusiasm over the technical triumphs of the energy transition to grave warnings about the size of the task that lay ahead.

On the first day of BETD, the CEO of Siemens—the massive German engineering firm that had put its entire global corporate weight behind building a low-carbon future—told the assembled dignitaries "the coming decade will definitely be the most dynamic decade in history" for the energy business, an "energy revolution" already under way around the world. At a session on financing that tectonic shift, a venture capitalist from Breakthrough Energy Ventures, Bill Gates's clean energy investment firm, asserted that we were gathered "at the dawn of a global transition to a clean economy." In a wonky session late one afternoon, one of the architects of Germany's own *Energiewende*, Hans-Josef Fell, ran through the highlights of his think tank's new report, laying it out in the clear technical language of the physics teacher he once was, explaining that powering the entire world with renewable energy by 2050 was now technically feasible. Fell had been a town councillor for a Bavarian burg of 11,000 souls when he pioneered the legislation that would turn Germany into the first industrial powerhouse of the clean economy. He knew intimately what success against stacked odds looked like.

At a side conference for journalists, I listened with particular care to an analysis by Kingsmill Bond of the Carbon Tracker Initiative, which had risen in a few short years from a sort of fringe shareholders' advocacy group advising London bankers on the risk of coal company stocks to an international force advising the world's financiers on the dangers of "stranded assets" (a term they coined for fossil fuels that will never be burned as the world shifts to a low-carbon economy). Bond confidently declared that little of what European energy planners were doing even mattered much anymore; China would set the pace and scale of the global energy transition, and it appeared to be fixed on faster and bigger.

But for each bold pronouncement of emergent success, there was a darker counterpoint. A spokesperson from the German Foreign Office worried whether the rest of the world was keeping up with its homework. One senior official after another from one country after another confessed that the pace was not yet quick enough, the targets not yet fully in sight. On the first morning of BETD, German foreign minister Heiko Maas had kicked off the proceedings with a quote from Greta Thunberg: "Change is on the horizon, but to see that change we also have to change ourselves." This is a reworking of Gandhi's oft-quoted maxim about being the change you wish to see in the world, and in a forum like BETD, it's a way of admitting that world-changing ambitions often reach beyond what any one of us can possibly hope to accomplish. Invoking the urgent promise of the student strikes in that conference hall served to underscore not the presence of action on the scale being demanded but rather its absence, to acknowledge the gap between what everyone understood to be the magnitude of the task and what the political institutions represented in the conference hall were capable of achieving.

At a press conference, Maas and his colleague, energy minister Peter Altmaier, took a question from a German TV news reporter. The reporter wondered, given the ever more urgent scientific reports on the crisis, why the German government wasn't doing more and moving faster. The answers of both ministers were given in a gently defensive tone best understood as the Esperanto of high-level climate politics, each in turn attesting to all that Germany had done and planned to do while outlining all the complex and varied challenges of it all.

At an award event one evening for start-up companies developing energy transition technologies, I ran into a colleague from Canada. The ceremonies had taken over a huge old warehouse in the port district of Berlin. The vibe was very digital economy—old beams and exposed brick, expensive fusion bar food, companies with cleverly vague names like Enapter (hydrogen-powered generators) and Blixt (smart grids). We spent most of the evening wondering aloud whether the grand project being celebrated around us was sufficient to stave off catastrophe. I've had a thousand conversations like this, and they all shake down to the dark euphoric two-step:

You have to admit it's not nothing. It's all come a long way, really roaring along now, right?

Sure, but I mean almost no one's on track for their Paris targets, and even they aren't enough . . .

And then we found another drink and chatted and shrugged in the evening-gala version of the ministers and their climate politics Esperanto. I'd heard the same message as well in the sub-text of a dozen German officials I'd interviewed over the years:

> Mein Gott, do you have any idea how hard it is to shift an industrial economy running at full speed from one energy regime to another? All at once, in a hurry, against the will of any number of established business and political power bases? With the cost immediate and significant, the risk still seen as distant on most horizons, and the rewards distributed unequally on a largely unknown timeline?

I tried—and failed—to imagine a war footing emerging from such talk. What would the precipitating event look like? What climatic Polish invasion or ecological Pearl Harbor—greater than the wildfires, floods, droughts and hurricanes already ravaging pockets of the planet daily—would compel everyone in that hall and beyond to drop all reservations and abandon all fossil-fuelled prosperity to join the Allied cause? What political tool would suddenly materialize to allow them to circumvent the march they were all on, that stumbling slog up a trail riven with obstacles, opponents, uncertainties, inertia?

Even a smaller community of the world's great powers united as the vanguard of an Allied effort proved impossible for me to conjure up convincingly. If the United States, the European Union, Japan and Canada—four of the world's ten biggest polluters, allied already in everything from the OECD and G7 to the IMF and WTO—joined together on a war footing, would that not be a substantial force to combat climate change? Surely it would. But as we gathered in the old Berlin warehouse that evening, the British were hell-bent on quitting the European Union and the American government even more determined to leave reality itself behind. It was easier to imagine cold fusion than anything like the historic summits of Roosevelt, Churchill and Stalin.

And yet climate activists have continued to wait on that one decisive moment for as long as I've been paying attention. My social media feeds seem some days like a convention of extreme-weather junkies, each vying with the other to make the most breathless post about record-breaking heat in France, or Indian cities bereft of water, or the post-apocalyptic cataclysm of California's wildfires. More than a decade after Hurricane Katrina turned half of America's Gulf Coast into a livestreamed dystopian movie about climate refugees, I find myself deeply skeptical that some incontrovertible inflection point is about to arrive.

Besides, the most profound lesson I ever learned in Berlin suggests that a war footing isn't the most efficient way to catalyze the global energy transition anyway.

As an Aside: On Forecasting

Angela Merkel, later to become the "climate chancellor," described one of the earliest goals of Germany's *Energiewende*—to generate 20 percent of the electricity on the country's grid from renewable sources by 2020—as "unrealistic." She was in opposition at the time, in 2005. Germany raced past that target in 2011. As Merkel prepared to step down as chancellor in 2021, about 43 percent of Germany's electricity was coming from renewables.

Merkel has nothing, though, on the IEA, whose record of repeated skepticism in its estimates of the growth of solar power over the past twenty years stands as a singular monument to the folly of forecasting. In 2002, for example, the agency estimated that 18 gigawatts of solar power would be installed worldwide by 2020. The world's grids saw 18 gigawatts of solar power added in 2007 alone.

In 2010 the IEA estimated that there would be 180 gigawatts of solar worldwide by 2024. The world's grids surpassed that estimate four years later, around 2014, and saw another 127 gigawatts added in 2020 alone. Also in 2010, the agency predicted that solar power would cost about twenty-two cents per kilowatt-hour in 2020, by which time the actual world average price was closer to five cents.

In 2016 the IEA predicted that additional annual solar installations would plateau from then on, at around 50 gigawatts per year through 2040. As already noted, 127 gigawatts of solar power was installed worldwide in 2020. It goes on like this year after year, with only slight improvements.

Point being: few saw cheap, ubiquitous solar power coming, least of all the established expert analysts of the conventional energy industry. They're not much more likely to be accurately gauging the energy transition's potential now. Assume that it's moving faster than anyone anticipates.

2.8 Epiphany at the Café Einstein

The first time I visited Berlin was at Christmas in 1992. My parents lived on one of Canada's last remaining military bases in Germany, and I was midway through my first year of university. We took a family trip to see newly unified Berlin, driving across the former East Germany and back. In the small towns and cities we saw Communist posters and Lenin memorials not yet torn down. Dresden still bore scars from the firebombing that razed it nearly to the ground during the Second World War.

Berlin was a strange city caught between two countries, two ideologies, two eras. Great stretches of the Wall still stood, caked in layers of graffiti and pockmarked where souvenir hunters had chipped away shards. At the Brandenburg Gate, old-timer veterans of the East German military were selling pieces of the Wall mounted in cheap Lucite, lining up the trinkets on blankets alongside old medals and fake-fur Soviet army hats. Not far from Alexanderplatz, with its iconic communications tower marking the heart of former East Berlin, we found a traditional *Gasthaus* for dinner one night, and the friendly server seemed mystified that anyone from glitzy North America would choose to spend a holiday amid her collapsed civilization.

On our last evening in the city, my brother and I left our parents at our downtown hotel and lost ourselves in the squatter communities that had taken over East Berlin's many abandoned warehouses. We got giddy drunk on cheap beer and the electric energy of a city reborn, high on gleeful anarchy, not to mention the joints that passed from table to table in the makeshift cafés run by the artists' collectives occupying the old warehouses. No one seemed to know what the rules were, or if there were any.

Few cities anywhere on earth have had the chance to reboot nearly from scratch this way, in the full ecstasy of liberation. I left thinking of Berlin as a maelstrom of creative chaos. After several more visits in the quarter century since, I've seen nothing to dissuade me from that view. Berlin is one of the world's pre-eminent cities for the art and practice of navigating historic change at a chaotic pace and civilizational scale—the very centre of the kind of epochal shift the climate crisis compels, the *Energiewende*'s living laboratory.

I went back to Berlin in 2006, not yet fully aware of how central it was becoming to the nascent energy transition. I was new to the climate-solutions beat, and I was on my way to Freiburg, in the far southwest of Germany, where I'd heard a revolution in solar power was under way. That story checked out, but it was only after talking to the folks in Freiburg that I understood that the catalyst for it all was a legislative coup years before in the nation's capital—a powerfully transformative package of legislation that made Germany the most welcoming home on earth for renewable energy. The solar panels might have been in Freiburg, but the motive force for the transformation was in Berlin.

When I next returned, in 2009, I understood much more clearly where I was—at the centre of what I hoped was the start of a global wave of climate solutions. Let me be clear about where climate solutions stood around then. The fall of 2009 might not seem that long ago, but the disposition of the world's political and economic elites toward renewable energy and other cleantech innovations was still predominantly one of abiding skepticism. The big money remained mostly elsewhere, and the epicentre of the global energy industry was still fossil fuels. Investment capital was pouring into the exploration and development of Alberta's oil sands and fracked shale oil across the United States and deep-water drilling from the Gulf of Mexico to the coast of Brazil. China was expanding its coal production so rapidly that a pervasive and only half-true myth emerged that it was adding a new coal-fired power plant to its grid every week. (In truth, Chinese coal consumption would begin to plateau in 2014, though it had indeed reached a formidable level by then after a decade of rapid expansion.) Natural gas had begun to position itself as a "bridge" to a greener future—a future still broadly understood to be many decades off.

I'd spent half of 2008 doing field research on the solar industry, and I repeatedly encountered the conventional wisdom that solar panels, as enticing and virtuous as they might be, would remain a bit player in the energy game for a very long time yet. Even in the solar industry itself, many companies were betting on next-generation materials and technologies, convinced that conventional photo-voltaic solar cells assembled from silicon wafers would never be efficient enough to compete in the big leagues. The massive economic stimulus package brought in by President Obama in the wake of the global banking collapse—its strongest legacy possibly the investment boost that made shale-fracking technology viable—had also bet big on some clean energy initiatives, triggering a wave of groans on many fronts. (The collapse of a solar company called Solyndra two years later seemed to confirm the wisdom of the skeptics, prompting the *New York Times* editorial board to clarify that Solyndra's failure didn't mean the end of the "suppos-edly doomed solar industry.")

All of which is to say that, just a decade or so back, there were precious few signs that the director of the IEA would, by the time it released its 2020 *World Energy Outlook*, be proclaiming solar "the new king of the world's electricity markets" and endorsing a renewably powered future as the inevitable way forward. This was the context for the most important interview I have yet con-ducted on the climate-solutions beat.

Here's the scene: Berlin, September 2009, early evening and warm. A patio table at a small, elegant café called Einstein on Unter den Linden, across the street from the office block where German parliamentarians have their workspaces. I was sitting with Hermann Scheer, a sixty-five-year-old member of Germany's Bundestag, who had been representing a district in the southwest state of Baden-Württemberg on behalf of the Social Democratic Party for twenty-nine years. Scheer was the co-author and champion of Germany's pioneering renewable energy legislation, which nine years earlier had kicked the country's fledgling efforts to reduce the greenhouse gas emissions from its electricity grid into roaring overdrive. I was hoping Scheer would tell me a little about how he'd launched Germany's *Energiewende*. Instead he provided a concise primer on how to catalyze an energy transition worldwide.

The *Energiewende*, in brief, went like this: In a small town in Bavaria in the early 1990s, a local councillor—Hans-Josef Fell, who I would later encounter arguing for 100 percent renewables at the Berlin Energy Transition Dialogue—had come up with a policy instrument to encourage local homeowners to put solar panels on their roofs. It was called a "feed-in tariff," and it obliged the local electricity utility to purchase power generated by the panels at rates well above those for conventional power. This essentially turned the value of reducing emissions into a cash reward for the solar panels' owners, making the pricey panels worth the investment. The technique worked, triggering local solar investments and eventually sending Fell to parliament as a Green Party MP when Germany elected its "Red-Green" coalition in 1998.

The coalition brought together Fell's Greens and Hermann Scheer's Social Democratic "Reds." And although both MPs were backbenchers, together they convinced their caucuses that a strong

incentive for renewable energy would be an interesting new direction for Germany's economy, which was still struggling to find its
stride after the shock of reunification. In 2000 the Red-Green
government passed the Renewable Energy Sources Act, a national-
scale feed-in tariff. And since reunification involved the particular challenge of absorbing an entire collapsed economy, the fact
that some of the manufacturing of solar panels and wind turbines
and other clean technologies could put East German workers
back to work in retooled factories was seen as an added bonus. In
particular, a shuttered Soviet-era production hub for photographic
film and other such products, located in a derelict industrial cluster
a hundred-odd kilometres south of Berlin, proved to have tools,
resources and skilled workers that could be readily repurposed
for making solar panels.

Within a few years of the introduction of the national feed-in
tariff, German engineering expertise and investment capital had
rapidly rationalized the solar industry. German firms jumped into
the front ranks of the wind industry, and the nation's manufacturers, designers and engineers became world leaders in everything
from green architecture and smart-grid components to the manufacture of hyper-efficient building materials and the application of
sustainable design principles to fashion and media and marketing.
By the time Angela Merkel's conservative Christian Democratic
Union toppled the Red-Green coalition in 2005, the *Energiewende*
was too far under way to abandon, even though Merkel had not
supported Scheer's strategy while in opposition. There were simply
too many Germans working in this new clean energy economy and
benefiting from solar panels on their roofs or the local power company's profitable wind farms to switch direction.

Merkel eventually slowed the pace and scaled back the generous
rewards provided by the feed-in tariff. But Germany continued to

play a pacesetting role in the global shift, even as fervent opposition to nuclear power, renewed by the Fukushima disaster in 2011, obliged her conservative government to phase out the nation's nuclear fleet and continue to rely on coal power, reducing the *Energiewende*'s impact. (In 2020, Germany finally announced plans to eliminate coal by 2038, though most analysts expect the job to be completed much sooner than that.)

In opposition, meanwhile, Scheer became a global champion of the German approach to climate solutions. He helped disseminate the feed-in tariff idea to dozens of other jurisdictions. He pushed for creation of the International Renewable Energy Agency (IRENA), an industry advocacy group meant to counterbalance the armies of fossil fuel lobbyists, and then let someone else run it so European pacesetters wouldn't dominate it. By the time I met him on that Berlin café patio in 2009, he was already a powerful global advocate for the energy transition, legendary in certain wonky energy-policy circles.

The foundational moment for Scheer as a climate-solutions pioneer was the German government's vote on ratifying the Kyoto Protocol in 2002. He was sharply critical of the approach, and when the Bundestag voted to join the European Union's first emissions trading system two years later, only a single dissenting vote was cast against the policy—by Hermann Scheer. The logic of that vote is crucial to understanding the scope of his vision for climate action and the pugnacious audacity of his political strategy. Here's how he explained it to me that evening at the café:

> The problem of these whole Kyoto Protocol climate negotiations is that they estimate steps to overcome these emissions by a shift to renewable energies—this is estimated to be an economic burden. And based on this premise, they come

automatically to the burden-sharing bazaar. Automatically. And this at a global level, with countries which have very, very different economic developments. Hmm? And very, very different energy consumptions. If you would recognize that this is not an economic burden, but this creates a lot of new benefits, including economic benefits, nobody would need the treaty.

What Scheer meant was that the UN-led emissions treaty process assessed climate change as an environmental problem to be regulated, thereby representing a drag on the global economy. It was blind by design to the opportunities the transition presented. After all, the IPCC, which set the parameters for the process, was a team of scientists, not entrepreneurs or investors or economists. But Scheer had known from the moment he first grasped the scope of the climate crisis in the early 1990s that the only viable solution was to supplant the existing energy industry with a new one that didn't burn fossil fuels. And so he assessed the problem as one of changing the industrial basis of modern technocratic society—ultimately, of building a much better energy system— not making pledges to accept a share of the cost of doing less bad. His no vote on the emissions trading pact was a defiant symbol of this commitment. Scheer wouldn't waste his time at the burden-sharing bazaar, arguing about whose sins obliged them to carry the greatest loads. He would dedicate his efforts instead to laying the foundations for that new industrial order.

Highly skilled politicians, especially those with a few big wins under their belts, have a way of explaining their choices and plans that makes them sound inevitable. Even speaking in his second language after a long day of parliamentary arguments, Scheer radiated that kind of confidence. He wasn't trying to persuade me. He was explaining how it was, how it had gone and how it would go.

An old leftie, he indulged in a story from the Cuban Revolution as a metaphor. Fidel Castro and Che Guevara, he told me, had used a "focal strategy" to conquer the Cuban countryside, village by village and farm by farm.

> The idea was, we conquer one region and from there we take the next. And in each region, we start immediately to change things. Immediately they expropriated the land of the big landlords, immediately gave it to the farmers. So people could see they work for us. Hmm? They work for us. And therefore they got more and more support by the normal people, and more and more people wanted to contribute in the fight—young people. And then it lasted less than two years, and then they reached Havana—and took it over. Hmm?

Implicit in Scheer's focal strategy was one crucial principle: the energy transition would not attack the established authority directly. From his seat on a back bench in the Bundestag, Scheer did not launch a foolish war on coal mining or an activist campaign to stop the construction of new coal plants. He organized a kind of dispersed insurrection, a guerilla-like effort, and that meant he needed allies. So he and his colleagues brought in a "100,000 Solar Roofs" program, subsidizing the cost of solar panels for 100,000 German households. To stodgy old German energy bureaucrats and big centralized power companies, that seemed like little more than a marginal indulgence and a negligible expense. To Scheer it was 100,000 German households (and their envious neighbours) already supporting the *Energiewende* before it reached full flight with the national feed-in tariff. Ahead of the tariff's introduction, he also found lawyers who were ready to make the case before international trade courts that it didn't violate trade agreements or otherwise offer an easy target for the lawyers of powerful incumbent utility companies. By the time the conventional energy industry in Germany knew it was under full-scale attack, the battle would mostly be won.

Roof by roof, town by town, the *Energiewende* built its support. Electricity costs rose, but Germans also saw opportunities to generate and sell their own power and reduce their energy use, offsetting the spike in prices per kilowatt-hour. They saw new jobs emerge in the renewable energy business, and they saw their local energy utilities making money from a wind farm or large-scale solar array. They even discovered a sort of national duty in it, one that stood in sharp contrast to the dark shadow that Germany's ambitions had cast over the twentieth century.

The day before I met with Hermann Scheer in Berlin, I spent the afternoon talking to officials at the Germany Trade and Invest office, a federal agency that was overseeing much of the energy transition at the time. One of them, a solar industry specialist by the name of Tobias Homann, was addressing rising energy prices. "The German electricity consumer pays a higher price for this energy revolution than others do," he told me. "And so people, I think, will thank Germany in the future for its role." There was a hint of pride in his voice. This was historic. It would matter to the whole world.

For all these reasons and more, the German public has remained strongly in support of the transition—often more than 90 percent say they agree with the shift, which is virtual unanimity in a modern democracy—even as costs have sometimes risen and hurdles emerged. There was something irresistible about the shift, something inevitable.

After we finished our interview, Scheer and I stood and shook hands, and then he scooped a half-empty bottle of sparkling water off the table, grasping it between his fingers by the neck. He walked off into the gentle dusk with the bottle dangling loosely from his left hand, like a cartoon Lothario carrying wine to a certain seduction. Scheer passed away from a heart attack barely a year later, and this was the image that came to me when I heard the news: a man strolling down a Berlin sidewalk through long shadows, somehow easy and loose in step despite his husky older man's gait. Carefree and oozing confidence. Certain that he'd already changed the world for the better.

2.9 Political Will Is Not the Easy Part

Hans-Josef Fell knows more intimately than most how to marshal complicated climate solutions from the margin to the mainstream. After more than a decade watching the feed-in tariff policy he'd nurtured and championed transform Germany's energy industries, he left politics in 2013. His victories as a parliamentarian were formidable: Fell's efforts, working as the Green Party's *Energiewende* co-champion alongside Scheer, were instrumental not just in putting solar panels on German roofs but also in making renewable energy viable as a mainstream power source around the world. As energy transition expert Ramez Naam once put it, "The most effective climate policy of all time . . . is Germany's early subsidy for solar and wind. These policies had impact not because of the emissions avoided in Germany (relatively small), but because they made solar and wind cheaper for the world."

But there was still a skeptical world of power brokers to sell on the larger goal—not just ubiquity but dominance. And so Fell founded a think tank called Energy Watch Group, and he went to the BETD in 2019 to present the details of a report titled *Global Energy System Based on 100% Renewable Energy*. The report seized on a slogan popular among the latest wave of climate activists—"100 percent is possible"—and gave it real technocratic teeth. Fell and his team had crunched the numbers and laid out a road map to 2050 in which the whole planet ran on renewable energy. Nearly 70 percent of electricity would come from solar, another 18 percent or so from wind, and the rest from some legacy hydropower and renewable biomass sources, all at costs not out of line with current prices. Transport would be powered predominantly by electricity and hydrogen fuel. Some mix of all of that would power heavy industries and heat smelters and otherwise keep industrial society intact. Existing policy mechanisms—Fell's own feed-in tariff, plus carbon pricing and utility auctions and the abolition of subsidies for fossil fuels—could make it all economically feasible. The target was fully in sight.

Fell is not an animated orator, but the sense of possibility in the room that day in Berlin was palpable. In countless other sessions at BETD, the talk spun around limits and challenges and the deepening crisis. Now here was one of the architects of the *Energiewende*, the very reason why this particular crowd was gathering in this particular city to have these conversations, telling us all that no technical barriers remained to an emissions-free future.

Did I mention that it was not one of the larger rooms in the venue? Not one of the biggest crowds? There's a post-reunification extension to Germany's Foreign Office building, stylish but not particularly grand, and a few dozen of us were gathered in a sort of library space there. The crowd trended younger and less stuffily dressed than the BETD norm. It felt like the vanguard of the vanguard, which is almost a way of saying the fringe of the fringe. I've seen several other presentations and lectures on the viability of a world fuelled only by renewables, though never one by a battle-toughened veteran of the climate fight like Hans-Josef Fell. Still, as always, I was waiting on the crucial piece of the plan: *How?*

"The energy transition," Fell said, "is not a question of technical feasibility or economic viability. It is only a question of political will."

There it was. Only political will. *Only* political will? Well, how hard could that be?

Here again is David Roberts, who has thought about these questions as thoroughly as anyone I know, writing at Vox on the subject: "Political will is not some final item on the grocery list to be checked off once everything else is in the cart. It is everything. None of the rest of it, none of the available policies and technologies, mean anything without it. It can't be avoided, short-circuited, or wished away."

Roberts uses the parallel example of global poverty to hammer home the point. There are no technical hurdles to solving it, no economic impediments in a world of trillion-dollar defence budgets. I'd add that poverty could likely be solved in less than a year. Bank transfers and disbursements are nowhere near as complicated and time-consuming, after all, as manufacturing and mounting enough solar panels to provide nearly three-quarters of the world's electricity. Why, then, hasn't poverty been solved?

Because political will is not the easy part. As Roberts says, it's *everything.*

2.10 Climate Politics 101 (Or: How I Learned to Stop Worrying and Love Politics and Then Hate Politics and Then Realize I Needed Politics No Matter How I Felt about It)

I am not a joiner. Growing up, I was not the overachieving kid who everyone in class said would be prime minister one day. I never wanted to organize the fundraiser or attend the council meeting. But after a few years on the climate-solutions beat, I felt compelled to try to use the experiences I'd had to do *something* in my own backyard.

I was still fairly new to Calgary, which at the time was a rapidly expanding boomtown drunk on oil-and-gas money, and which had been run in the civic and community sense mostly by a fairly small network of old boys (and far fewer girls). But it was also growing fast, its demographics changing all but overnight as thousands of newcomers arrived each month. And it prided itself on its social fluidity, its openness to new people and new ideas. Calgary felt like a city ready to be changed. But what to do?

I lived in an old neighbourhood on the edge of downtown, tucked beneath a steep ridge known as Scotsman's Hill (allegedly in tribute to the legendary frugality of the Scots, because you could watch the Stampede's rodeo events for free from its crest). There was a proposal for a large, mixed-use development in my neighbourhood, a mere three blocks from my house. It was the sort of multi-tower, multi-function construction project I'd been seeing in forward-thinking cities around the world, the ones held up as models of sustainable urban design. There would be residential and office space on the same site, and it would include retail and other amenities that would serve the light rail transit (LRT) station that would be built nearby in the years to come.

Urban density! Mixed-use development! Transit-oriented design! The project was torn from the pages of my own writing on what cities needed to do to prepare for the energy transition. I decided to attend the next community association meeting, voice my support and lend whatever knowledge I could to the proceedings.

And that's how I became one of the least popular regulars at the monthly meetings of my local community association. This was one of my first lessons in politics, and among the most important ones I've learned. In fact, I'd recommend that anyone with big ideas for faster, better, more climate solutions attend the nearest community meeting they can find, in order to hear how people respond to proposals to bring some initiative aimed at the greater good to their backyard. It's not that no one wants it. It's that the ones who *really* don't want it are far more motivated to organize and take aggressive action than the ones who think it would be fine.

In my case, every community association meeting I attended on the topic dissolved into the protracted, multivalent, incoherent debate and rhetorical crossfire I've since learned is endemic to such proceedings. Again, it's not that everyone hates the idea, but few are as ferociously dedicated to grinding the process to a halt as the ones who oppose change. (They have a nickname, of course—they're NIMBYs, because their response is always "not in my backyard.") As one of the more prominent voices in favour of the new development, I earned the ire of several long-serving members of the community association's executive. One evening before a meeting, one of them (who had somehow found my phone number) called me to suggest that I shouldn't attend unless

I wanted to embarrass myself further. I never did get clear on what was embarrassing about voicing support for transit-oriented development, but presumably I'd been cast by the NIMBYs as a dupe of Big Condo.

When the proposal appeared before city council for community input, a group of us went to the meeting to argue in favour of it. What sticks most with me about that day was that several opponents of the development had figured out a clever way to argue against the alleged green features of the condo and office buildings, by arguing that the new towers would cast shadows over their roofs—which, they claimed with straight faces, they might one day want to cover with solar panels.

In the end, a downturn in the local economy killed the project. I'm relating it in detail because the lesson remains so important. The better idea—the more necessary one, the one that pointed directly at reversing the car dependency that was deepening the climate crisis—was not enough to convince a community to change. Even then, more than ten years ago, I was well aware that the technical barriers to tackling climate change were vanishing fast. But I was also beginning to realize that the best ideas are not always (or even often) the ones that carry the day. It was not enough to be right. You had to find a way to win, and that meant winning people over to the value *for them* in your idea.

I'll repeat David Roberts's crucial line: political will is everything. And political will inevitably involves with the extraordinarily complex task of engaging people who might not care about your big ideas or even be openly hostile to them. In another piece for Vox, Roberts dug into what the elusive concept of political will really means, summarizing a 2010 study by researchers at Yale and South Dakota State universities. The study defined political will as "the extent of committed support among key decision-makers for a particular policy solution to a particular problem," breaking down the idea into three essential components. The first is "the distribution of preferences"—the range of different priorities in a particular democratic population, the question of *who wants what change right now.* The second is "the authority, capacity, and legitimacy of key decision-makers or reformers"—the question of *whether those who want that change have the power to make it happen.* And the third is "commitment to preferences"—the question of *how much those in power want that change.*

Compare these criteria to the usual methods of advocating for action on the climate crisis, and it becomes clear why change has been so maddeningly slow at the political level. Does a protest march express deep commitment to one preference, one action? It very often does, muddy as the messages on competing placards can be at such events. But do the marchers have the power to force that change? Rarely—a march, after all, is an attempt to persuade the powerful to embrace the issue—and when they do, it's usually only after years and years of slogging progress: building a movement, finding allies in government, moving an issue from the bottom of a political party's agenda to somewhere near the top, and then waiting for that party to gain the power to enact the change.

Climate activists have sometimes looked to the civil rights movement for inspiration, but in my experience they carry a cinematic notion of the change it brought about, a few quick cuts from the streets of Birmingham to the March on Washington to the signing of the Civil Rights Act. This version edits out the decade that elapsed between the Montgomery bus boycott and the Act's passage, and the years yawning into decades afterward, when Martin Luther King and his successors carried on the fight in the wake of only a partial victory. John Lewis, among King's most successful acolytes, was still pushing in Congress for progress in pursuit of the movement's incomplete goals when he died in July 2020.

You want change? This is what it looks like. Messy and maddening, compromised and contradictory, led (often as not) by dishonest, self-interested brokers who are in command (often as not) of political systems and institutions designed to resist change. And there's no skipping past the politics. There's no magic lever that, once pulled, will simply grant war-footing-like command and control, allowing the best ideas to be transformed into policy and enacted with frictionless ease by a populace that has granted full consent to the program for as long as it takes. One by one, climate solutions have to pass through the same grinding gears as the one mixed-use development someone once tried to build in my neighbourhood.

The next chapter of my own political education was at citywide scale. I'd decided that community associations were not vehicles for rapid change, at least in my end of town, so I joined the board of directors of a civic group called Sustainable Calgary. (As I recall, I found it by googling the words "sustainable" and "Calgary.") It was a small, academically minded organization that issued reports on the state of the city according to a range of sustainability metrics (greenhouse gas emissions, clean energy and transit use, that sort of thing) and then did what it could to urge the city to do better on those metrics. It was an earnest and well-meaning group, but I found it lacking against my emerging litmus test of *How?*

I'd heard from colleagues in other civic-minded groups that simply getting people to show up at city hall to push for change could be maddeningly difficult. I'd recently attended a tech-industry function known as a BarCamp, a kind of one-day conference built on the novel idea that the attendees decided at the start of the day what the agenda would be. So I joined a few of those colleagues in organizing a one-day engagement event on municipal politics, which we called CivicCamp and ran loosely on the BarCamp model. One bright idea that came out of it was to get volunteers to sit in on council sessions on behalf of the many people who'd like to participate but couldn't wait through most of a day's agenda for their item to come to the top of the list. CivicCamp volunteers would notify the interested parties when their time was about to come up, so they could duck out of work and have their say.

It didn't feel like much, but we heard from city councillors that our pressure looked unprecedented from their end. Fast-forward a year or so, and one of the original CivicCamp organizers declared his intention to run for mayor. This struck me as almost absurdly audacious. I thought he might have an outside shot at a council seat, sure, but mayor? Still, my wife and I ordered up a lawn sign and helped a little with his campaign, which we thought was doomed to fail right up to the moment he was declared mayor of Calgary. By now you may have heard of him—his name is Naheed Nenshi.

Did Mayor Nenshi become Canada's greatest climate champion? Did he transform Canada's oil capital into an energy transition hub? Of course not. He governed smartly, often strongly, by increments. He won some and lost some. But he understood the value of urban density and more transit and preparing a city for the dramatic changes to come. In the planetary scheme of things, it was the tiniest of victories. We celebrated wildly on the night of his first electoral win, then less so the two election wins after that, as the grind of running a city in which he had merely one vote of fifteen in a council burdened with wildly differing views took its toll.

What happened next in my political education still sounds preposterous in the retelling. I ran for Parliament in a federal by-election—as a Green Party candidate. In downtown Calgary, the nation's oil capital, which at the time was home to the sitting Conservative prime minister. That was one facet of the absurdity: imagining that oil-rich Cowtown would somehow elect a Green candidate. The other part was that there was little in my personality and skill set to suggest I'd be well-suited to being a door-knocking, glad-handing political candidate. But I ran for Parliament anyway, because smarter political minds than mine (including several veterans of Mayor Nenshi's first victorious campaign) thought I actually had a shot, though an extremely slim one, at stealing the seat. That was the hook for me. I was willing to chase even the slightest of possibilities that I could spend the next few years in Ottawa, seated across from the prime minister who had pulled Canada out of the Kyoto treaty, declared war on climate activists, and sneered repeatedly at the very idea of climate action, representing his city.

What I didn't anticipate was how much I would learn as a political candidate. How much the absolutely awe-inspiring volunteers, many of them veterans of multiple political campaigns who treated it as their unpaid second job, would teach me about how politics works at the level of the street corner, the community hall debate, the voter's doorstep. How much the voters themselves taught me, in very short order, about the vast chasm between my issues and their daily needs and deepest concerns, and how my priorities were as important as theirs only if I could somehow make an unequivocal case for that fact. How much I would learn from the seasoned campaign pros about the messy brawl of party politics, the way it amplifies small disputes and exaggerates almost non-existent distinctions, and, more than all that, how little those esoteric rhetorical exercises matter compared to the actual goal of politics, which is to acquire and wield power. That last lesson won't be pretty to many ears, and there are some in politics who will deny it's the main goal of the game, but I would argue that they're being either dishonest or delusional. There are a million ways to improve your community or advocate for an issue, but there is only one way to make new laws in a democracy—and that is to win elections.

I lost, of course. But my campaign was the most successful Green campaign ever in Alberta. The final tallies were 37 percent for the Conservative candidate, 33 percent for the Liberal and 26 percent for me—a very strong third, if such a thing can be claimed with a straight face. And there was a point, a couple of weeks before election day, when I came to understand where all the silly clichés that campaign veterans repeat about momentum and excitement and "what we're hearing at the doors" come from. It all seemed to be converging. We had hundreds of volunteers and glowing press and a big rally with national celebrities that felt like a rock concert and a poll that showed a three-way tie—until it vanished into the ether, as polls sometimes do, before it could be published. (Another of those hard political lessons: you can buy an unfavourable poll and make it disappear.) There were a couple of days when I actually had to consider from a logistics point of view, at least a little, how I would manage my life as the Member of Parliament for Calgary Centre—where I would live and where the kids would go to school and all that.

And then it was over and I was a private citizen again. I had entered the political fray because of the urgency of the climate crisis, a sense of the immediate need to act. I wondered if there might be better uses of my strongest skills than long-shot electoral politics. So I moved into the metaphorical backroom, helping wonky climate-policy advocates and energy-focused think tanks craft better messages and more readable reports.

But then, not too long after, another sort of offhand miracle occurred: the people who listened to people like me took over the government of the whole damn country. Let me assure you, I did not see *that* coming.

2.11 The Highly Qualified, Necessarily Compromised Thrill of Climate Victory

For a certain stratum of the Canadian policy-wonk community, as I noted earlier, the biannual Globe conference in Vancouver has become one of the highlights of the work calendar. I'm not entirely sure why—maybe because it's managed to stay agnostic enough that it attracts a broad spectrum of cleantech entrepreneurs, policy obsessives, elected politicians and climate NGO folks in equal measure. It bills itself as a "sustainable business summit and innovation showcase." The 2016 instalment of Globe, though, felt like something different. I came to think of it as Canada's climate prom.

That 2016 conference was the first opportunity for Canada's climate and energy policy crowd to gather in such numbers since the qualified triumph of the Paris climate summit at the end of 2015, as well as a pair of election victories in the months leading up to it that had stood Canadian climate politics on its collective head. The first was the shocking win in April of the left-leaning New Democratic Party under Rachel Notley in my home province of Alberta, toppling a Conservative dynasty that had ruled for forty-four years. In the heartland of Canada's oil-and-gas industry, a new government promising strong action on climate change now reigned. In the federal election six months later, the Liberals under Justin Trudeau ended nine years of government under Stephen Harper's Conservatives—a less surprising shift, but still a thrill for those who'd felt ground down by nearly a decade of brazen, often jeering inaction on climate change and relentless cheerleading for the fossil-fuelled status quo.

There had been a sort of weary fatalism to the climate beat in Canada in those years, a despair that hummed in the background even as significant regional victories accrued along the way (British Columbia has a carbon price! Ontario is shuttering coal plants!). But ten years can seem like forever in politics, so it felt as if the inertial weight of the Conservative government had hung over such gatherings for as long as they'd been happening. The glib references to "job-killing carbon taxes," the hostility toward all things clean and green, the home-team boosterism for oil and gas—it wore you down. On some level I think we'd internalized the idea that the collective project we were all working on would succeed, if it did, despite the federal government. We understood that we were outside, looking in.

I'm sure no one anywhere in the world who spent the decade before the Paris summit working on climate issues made it through without similar stress fractures in their psyches. There was no frictionless consensus anywhere. Germany fell back on its coal reserves when eliminating nuclear power became a political necessity. The vaunted green Norwegians still drilled for new oil in the North Sea, even as they carefully saved the profits for an oil-free future. Even my colleagues in Copenhagen, the urban cycling capital of the world, kvetched about intransigent local bureaucracies. China, meanwhile, couldn't build new coal plants fast enough, and the United States government had become a gridlocked partisan basket case. There were no unqualified victories.

At previous Globe conferences and other events like them, the triumphs were low-key, almost private. The headline speech would come from a mayor, or maybe a provincial minister. The business leaders who attended were a step or two below chief executive officer. The entrepreneurs were the sort who still had to talk about the marketplace in the future tense. There had never been a buzz of high office and wonky excitement like there was on that first morning of the 2016 conference.

The morning plenary was running a little late, as events often do when senior elected officials and their overpacked schedules are involved. I sat toward the back of a capacity crowd in the huge ballroom, watching little knots of half-familiar faces march in to find their reserved seats near the stage. Some of those faces I knew from the TV news—cabinet ministers, premiers—but others, exhilaratingly, were people I'd spent time with in those wonky sessions during the long Harper-era exile. There was the former director of an Alberta energy think tank, now chief of staff to the federal environment minister. Here came a former senior staffer for a range of environmental NGOs, whispering in the ear of the natural resources minister. I remembered an afternoon in Toronto not that long before, when we shared a cab to a suburban TV news studio to join a panel criticizing the disastrous climate stance of the government of the day. Now she *was* the government of the day.

It was hard not to be a little awestruck watching the rows of VIP seats fill with some of the smartest and most dedicated people I'd come to know on the Canadian climate-solutions beat, now fresh from the breakthrough at the Paris climate summit. Canada's delegation—which in the years of Conservative rule had been a perennial favourite for the "Fossil of the Day" award bestowed by environmental activists on the most obstructionist forces at the event—had helped lead the push to move the target warming threshold from 2°C down to 1.5°C. The gathering dignitaries were all in Vancouver that week because the federal and provincial governments were working out a coordinated policy strategy to commit the entire country to pursuing the climate goals embraced in Paris.

Eventually the steady flow of senior government officials stopped. There was a sort of lull as people found their seats and engaged in small talk that grew steadily more hushed. And then one of those mass flutters ran through the crowd as the final knot of VIPs entered the hall. I saw two familiar faces. The first was a colleague I'd met years ago in Toronto when he worked on a fledgling climate policy for the Ontario government. We'd crossed paths several times over the years since, but the encounter that came back to me as I sat at Globe that morning had occurred during the first few desultory years of Canada's decade-long Conservative turn.

My colleague had been passing through Calgary as part of his work as director of World Wildlife Fund Canada. We tipped back a few pints at my local pub, lamenting the maddeningly slow pace of action on climate change. He said something about how the WWF's most loyal and deep-pocketed donors tended to push back on climate campaigns, preferring the more familiar, less complex and less politically charged work on endangered species and photogenic habitats. And now he came striding into the hall, my old colleague Gerald Butts, shoulder to shoulder with the other instantly recognizable face in the group: Prime Minister Justin Trudeau, whom Butts served as general secretary. He had become an almost mythic figure since I'd last seen him, widely credited with orchestrating Trudeau's rise to the Liberal Party leadership and his victory in the 2015 election.

Trudeau's speech that morning at Globe served as a broad statement of intent about the energy transition then emerging.

The meat of it was a pledge of millions of dollars in new funding for clean energy, part of a larger promise to build Canada's future prosperity on the foundations of a low-carbon economy. It was received very well, not too far short of rapturously, by a crowd not used to such attention from a prime minister. I liked the speech quite a lot. But then, I would say that—I had helped write it.

There's an undeniable thrill in listening to the head of your country's government enunciate your words to a packed auditorium. (The bulk of the words, to be clear, were those of the Prime Minister's full-time communications people, but more than a few of mine made it through.) And the thrill was amplified, made even more surreal, by its improbability. I was not "in politics" in any formal sense. And anyway, whose first political speechwriting credit is for the prime minister? Whose first political speechwriting credit is for an address that aims straight at the treacherous political minefield of climate change and the energy transition? But there I was, anonymous to all around me in my near-the-back seat, an observer of my own writing as Trudeau attempted to navigate a country rich in fossil fuel resources toward a low-carbon future that only a small slice of the Canadian public outside the conference hall understood as a necessity.

Changes in government bring new priorities and new policies, sure, but a full about-face at such a pace is exceedingly uncommon. For a modern multi-party democracy governing a pluralistic, multicultural federation strewn across a vast land mass, the speech marked the political equivalent of an overnight 180-degree turn. And it was particularly striking on the climate change file, an issue almost wholly lacking in easy wins that could serve as victory laps in the next election campaign. There was one line in Trudeau's Globe speech that came directly at the messy complexity, and it was the line that reverberated most strongly beyond the conference hall.

"The choice between pipelines and wind turbines is a false one," he said. "We need both to reach our goal." The Canadian Press quoted the line at the top of its coverage of the speech, noting, "It was not an applause line in this green coastal city."

This was a shorthand reference to the defining conflict at the root of Canada's response to climate change, which is itself a microcosmic version of the whole world's conundrum. Canada is a country in which the majority of citizens had been telling pollsters for years before Trudeau's election that they wanted stronger action on climate change. But Canada is also a country with an abundance of fossil fuels, considerable reliance on the revenue and jobs they create, and in recent years a particularly strong symbolic and cultural attachment to the oil and gas produced in the western provinces. Canada is a country that was, when Trudeau spoke in 2016, nearly a decade into a ferocious debate over the construction of new oil-and-gas pipelines, especially those intended to carry bitumen from the oil sands of northern Alberta to world markets. And Canada is a country that elected a Liberal majority in 2015, in significant measure on its promise to bring in a national carbon price, incentivize the installation of more wind turbines and other clean energy technologies, and make real cuts to greenhouse gas emissions nationwide.

The political momentum in Canada had been pushing politicians and their supporters into one camp or the other with considerable force. More oil production, or emissions cuts? Drill, baby, drill, or keep it in the ground? Pipelines or turbines? Choose your side. And there was Trudeau calling for more of both. Some in the Globe crowd surely thought the pipeline reference was nonsense rationalization from a sellout politician, others that it was standard-issue pandering to both sides of the debate, a typical mushy Liberal centrist embrace of no stance at all. Those are understandable criticisms, but I don't think they're accurate. The pipelines-and-turbines line was, rather, a concise expression of the crux of the political calculation driving Trudeau's approach to climate change and the emerging energy transition.

Here's the calculation: Climate change is a long game, one of the longest that any government in Canada (or anywhere else) has ever had to try to play and win. Any effective action on the climate crisis must, by necessity, outlast the government of the day. A pledge to end coal or eliminate all carbon emissions by 2050 means nothing if it vanishes the moment an election is lost. And a government with real ambitions for climate change would surely hope to last more than one election cycle—to buy itself time for a carbon price to prove effective at cutting emissions while not crippling the economy, to show that big investments in renewable power and other clean technologies can generate opportunity and prosperity. For a Canadian government, the best shot at building a consensus on this approach is not to launch an all-out assault on the deep-pocketed fossil fuel industries, which employ hundreds of thousands of people and prop up the pensions and retirement plans of millions more.

Recall Hermann Scheer's tactical approach in Germany—not to topple the powerful German coal business but to dodge around it and build a successor industry. This, I would argue, was the Liberal government's political calculation, the logic behind the call for both pipelines and turbines that day in Vancouver. By avoiding outright war with the oil sands industry or any other arm of the fossil fuel business, the government hoped to buy time to embed its climate plan permanently in the halls of Canadian government, to make carbon pricing and cleantech and the end of coal and all the rest irreversible, to stitch declining emissions and an expanding low-carbon economy as fully into the nation's political fabric as universal health care. That, after all, is what all those holders of high political office were doing in Vancouver. The conference was a sidebar to the work of hammering out what came to be called the Pan-Canadian Framework on Clean Growth and Climate Change, a mostly successful effort to coordinate climate solutions nation-wide at both the federal and provincial levels.

Mark Jaccard, an energy economist at Vancouver's Simon Fraser University and often a sharp critic of Canada's climate policies, had this to say about Trudeau's political calculation: "Perhaps the government will build a new oil pipeline and will also miss its 2030 target. But these don't matter much for the global climate challenge. In climate policy, experts agree that Canada is finally a global leader." The whole world, Jaccard noted, is hunting for an equation that can accelerate the transition to clean energy while avoiding such jarring disruptions to the conventional energy economy that the consensus for action collapses.

This is what political will looks like. This is the whole game, playing for keeps, changing the bottom line on every company's balance sheet and recalculating line items on every tax return. This is ending coal within a generation, turning renewable energy into a primary power source, initiating a permanent shift away from internal combustion engines and the gasoline they burn.

Eight months later, Trudeau announced that his government was approving construction of the Trans Mountain Expansion project (TMX), a new pipeline to run alongside an existing one from just outside Edmonton to the port of Vancouver, intended to transport 800,000 barrels of bitumen per day. To many critics in climate advocacy circles, this was a betrayal of the Liberal climate plan, erasing entirely the promise of real progress and revoking Trudeau's climate credentials. (The federal government eventually bought the project to keep it from falling apart, and whenever federal Liberal officials post on social media about anything climate-related, their feeds rapidly fill with replies saying simply "You bought a pipeline"—considered sufficient rebuttal all on its own.)

This was the other side of Trudeau's political equation, and the process is not pretty. As debate raged about the TMX decision, Gerald Butts posted a response on Twitter: "If anyone has a better idea of how to implement strong climate policy in Canada, step right up." I believe he meant exactly that. It was the best idea they'd come up with to chart a new course for the sluggish old ship of state. It was as far and as fast as their political calculations told them they could go without losing the support they needed to keep governing. There was no special lever to pull to enact a war footing, no way to rapidly accelerate processes that require the unavoidable, often tiresome slog of corralling public support and building consensus.

This, I'd argue from my own on-the-job education, is the hardest lesson of politics. Nobody gets exactly what they want. You rarely get most of what you want, and so it's a win to get even some of what you want. You can spend a decade getting absolutely none of what you want. And that's why it feels like a real triumph to get even a highly qualified, necessarily compromised chunk of what you want served up to you in a Vancouver conference hall.

2.12 Memes Are Not Enough

I recognize the value of powerful slogans and principled stands. To create political will, people must be rallied to a cause, and these are effective tools for that purpose. Here's one prominent example from contemporary climate activism, which I mentioned earlier in passing: "Only 100 companies are responsible for more than 70 percent of the world's emissions." (Or, as it was more viscerally phrased by the left-wing news site Truthdig, "Just 100 Companies Will Sign Humanity's Death Warrant.") The meme alludes to a basic truth, which is that the majority of the world's greenhouse gas emissions can be traced to about a hundred companies and state entities involved in extracting, refining and distributing fossil fuels. It would seem to suggest the true enemy in the climate battle: a coterie of oil, gas and coal executives reaping profit from the devastation of the planet, while humanity suffers and struggles to break free of their clutches. For a diffuse protest movement trying to corral a political issue that encompasses virtually every nation, community and business on earth, the appeal of the meme is obvious. Here is an enemy force small enough to be rounded up, disarmed, defeated. Never mind the talk about complexity and compromise. Simply erase those hundred companies from the ledger and the task is all but complete.

Trouble is, the meme assumes not only a political program that doesn't exist but also an enemy political faction—the Hundred Entities—that exists only as a rhetorical flourish. The meme's origin, near as I've been able to track it down, is a 2017 report by a British environmental group called the Carbon Disclosure Project, which refers to the Hundred Entities as "carbon majors." The data seem to support the claim in a broad sense—the majority of the world's carbon dioxide emissions do begin their atmospheric journey in the industrial activities of a hundred companies and state entities. But the meme leaves out the most important information: many of the "companies" on the list are better known as nations, and the customers and users of the fuels produced by them constitute not an easily opposed political force but the vast majority of humanity. At the very top of the list, for example, is "China (coal)." As in the country, and its primary source of electricity.

China is not a company. It can't be coerced into ceasing electricity production. There's no political program extant anywhere on earth outside China that can change its levels of coal consumption. Lumping it in with, for example, Peabody Energy, a publicly traded American corporation with a board of directors who could vote (or be obliged by legislation) to cease operations tomorrow, is ignoring basic facts for the sake of rhetoric. "Only 100 companies are responsible for more than 70 percent of the world's emissions" is essentially a meme-friendly way of saying the global fossil fuel industry is a large-scale and highly centralized enterprise. It's a description, not a strategy of attack or a political plan of action.

There's nothing wrong with a great slogan or a powerful meme. But it's a misleading error to equate one with a full answer to that crucial question of how to create change. As I once read, in a review of Naomi Klein's climate action book *This Changes Everything* by Evergreen State College economics professor Peter Dorman, there is a real danger in "the treatment of goals as if they were programs." The danger, he argues, is that it represents "an adaptation to powerlessness."

"It's all ends and no means," Dorman wrote. "This is a double convenience: first it eliminates the need to be factual and analytical about programs, since announcing the goal is sufficient unto itself, and second, it evades the disconcerting problem of how to deal with the daunting political challenge of getting such programs (if they even exist) enacted and enforced."

To embrace the idea that climate change is the sole responsibility of the Hundred Entities is, in other words, to say that the question of solving it lies entirely in someone else's hands. The map, as the saying goes, is not the terrain, and the symbol is not an individual unit of the universal, replicated (or soon to be replicated) everywhere. The rules—policies and guidelines and targets, fuel standards and emissions limits, energy efficiency ratings and building codes—are the levers that really can transform the symbol into the new normal. A meme is worth very little unless and until it leads to control of those levers. And the first lesson such control tends to teach is that there is nothing simple about recalibrating those levers for big change.

As an Aside: The Denial Industrial Complex

The Hundred Entities might be a rhetorical convenience, but there has been a small, organized cadre of legitimate villains who have been working to actively derail climate action for as long as it's been a political issue. And they are connected to the fossil fuel industry. They are not corporations, though—they are unscrupulous scientists and the think tanks and public relations firms that have published and publicized their thirty-year campaign to cloud the science of climate change in doubt and denial.

In 1989, not even a year after NASA's James Hansen first testified about the emerging scientific realities of climate change in Congress, America's foremost polluters—including major American oil and coal companies, auto manufacturers, and chemical companies—came together to oppose climate action under the banner of the Global Climate Coalition. This and numerous other "astroturf" public interest groups received millions of dollars in funding, most significantly from ExxonMobil and the private fossil fuel company Koch Industries, to spread the word of a small handful of dissenting scientists, most of whom were not climate experts. Both the tactics and in some cases the personnel of this denial industrial complex trace their origins to campaigns funded by the tobacco industry to cast doubt on the links between smoking and cancer. By one estimate, ExxonMobil alone donated more than $37 million to climate denial groups from 1998 to 2019, in addition to the tens of millions the fossil fuel industry spends around the world each year to lobby directly against stronger climate policies.

There is no question that all this spending had significant impacts, particularly in the United States, where climate-denial groups and lobbyists played a crucial role in persuading the Republican Party to become by far the largest institutional supporter of fraudulent climate science on earth.

I have chosen not to dwell on the activities of the denial industrial complex in my discussion of the political barriers to climate action, for two reasons. First, I believe it is a mostly spent force outside Republican circles in the United States, and my focus is on the recent and future success of the global energy transition, which is occurring despite the tireless efforts of the deniers. And second, the record on climate action (and lack thereof) in parts of the world mostly beyond the reach of its propaganda campaigns indicates that foot-dragging on climate action is endemic to modern politics with or without the efforts of these "merchants of doubt" (to borrow the memorable title phrase of Naomi Oreskes and Erik M. Conway's exhaustive study of the movement).

The denial industrial complex fought very hard to halt climate action. It surely contributed to the many years of maddeningly weak climate ambitions among any number of governments and business sectors. But outside the American political arena, it has ultimately lost the battle.

2.13 Building Codes and the Necessary Process Grind

Climate solutions generally do not resemble the manifestation of a singular unwavering vision; they are, like any long-term political solution, especially one of such scale and complexity, about collaboration and compromise. They are about baselines, defaults, standards and regulations and codes. They are about process. And the process, though often dull and fussy and frustrating, is the only way to the goal. Which is a high-minded way of saying I'd like to explain how I've come to see great triumph in a lowly building code.

One way to tell the story of this triumphant building code is to begin with a one-off marvel of the sort that green advocates love to champion as the way forward: the Saskatchewan Conservation House. Rewind back to 1977, in the depths of the era's energy crisis, as governments around the world scrambled to find ways to conserve energy at the end of a generation that had operated giddily on abundant fossil fuels and declining energy prices. In Saskatchewan, the provincial research council commissioned an engineer named Harold Orr to construct a model of energy conservation in Regina, on the frigid Canadian prairie. The resulting house was a venerated wonder of renewable power, energy efficiency and zealous insulation, visited by thousands and studied internationally. And then energy prices plummeted, and Saskatchewan and much of the world forgot about the value of efficiency and conservation for another generation.

In the meantime, some of the core principles behind Saskatchewan's Conservation House jumped the pond to Germany, where a small townhouse block in the city of Darmstadt was built in 1990 to the hyper-efficient specs pioneered in Saskatchewan. German researchers and architects developed a set of standards that became known as *Passivhaus* design and established the Passivhaus Institut in Darmstadt to refine and disseminate the concept. In the early 2000s, North American researchers hunting for climate solutions rediscovered the value of energy efficiency, and Germany's *Passivhaus* hopped back over the Atlantic.

More than that, though, passive house design and other efficiency-obsessed ideas started to inform building codes. Consider the case of one of the world's most avowedly green cities, Vancouver, and the provincial building codes it operates under. Like any climate-conscious metropolis, Vancouver had by the early 2000s started to see Leadership in Energy and Environmental Design (LEED)-certified buildings and other hyper-efficient one-off construction projects, at around the same time passive house standards began making inroads in North American green building circles. Then in 2008, the government of British Columbia, already well ahead of the national and global curves with its carbon tax, began requiring local governments to incorporate climate targets and plans in their community planning and growth strategies. The municipalities lacked the tools to meet those targets, however, and the twenty-one local governments across Metro Vancouver wound up introducing incentives for any number of green building standards, resulting in a tangle of requirements that caused headaches for the construction industry. So industry leaders asked the BC government to come up with a one-size-fits-all approach. And thus did the slow grind of policy wonkery churn on, and in 2017 the final result was the BC Energy Step Code.

The new code leads the industry step-by-step to zero-emissions construction. Local governments aren't obliged to follow it right away, but all buildings in the province must meet those net-zero standards by 2032 one way or another—another one of those highly qualified climate victories.

And what a triumphant bureaucratic document it is, full of technical specifications and gradual step-by-step improvements, checklists and performance targets and compliance tools. On its face, it's all about as heart-racing as the phrase "provincial government working group." (Many years ago, the American political magazine *The New Republic* ran a contest inviting readers to suggest the most boring possible headline, inspired by one in the *New York Times* that read "Worthwhile Canadian Initiative." The BC Energy Step Code is nothing if not a worthwhile Canadian initiative.) And yet the dull language and process grind hide an achievement far greater than any *Passivhaus* or other architectural marvel. Under this building code, British Columbia is the first jurisdiction in North America to lay out rigorous binding requirements that ensure every new urban construction project in the province—every row of suburban houses, every apartment block, every office tower and warehouse—will be a wonder of energy efficiency by 2032. As good as a passive house. And as close to a fully realized climate solution as you'll find.

This is what climate victory looks like. There were no parades, no delirious celebrations under banners reading "WORTHWHILE CANADIAN CLIMATE INITIATIVE ENACTED!" It wasn't won in a day and it may not yield dramatic results for more than a decade yet. This is the necessary process grind, and its gears spin at speeds that can be increased only within limited ranges. One gear whirred to life in Saskatchewan as the OPEC energy crisis deepened, and in time that sent another one spinning among German engineers worrying over high energy prices there. The teeth of that one caught a few parts of a world beginning to hunt for climate solutions, and then Vancouver architects and policy wonks across the Salish Sea in Victoria locked together to spin faster still. The process started in 1977, sped up in 1990 and again in 2008, reached full speed in 2017, and will finish the job (so it's been decreed) in 2032.

The process grind of the global energy transition is accelerating. It looks less risky and more profitable by the day, even as the urgency created by the climate crisis becomes more self-evident. But acceleration is not the same as elimination, and there's no amount of evidence compelling enough to permit circumventing the process entirely.

2.14 The Science of Not Listening to the Science

When climate advocates call for stronger action and cite verified and legitimately terrifying facts and figures, when they post and repost memes from the latest natural disaster wrought by the climate crisis, they are targeting their actions at a broad public presumed to be composed of millions of mythical creatures known as "rational actors." The rational actor has been the star of more than a century's worth of economic theory, an animal presumed to be capable, when presented with clear information, of readily identifying its best interests and acting to protect and enhance those interests.

Belief in the existence of rationally acting masses can be readily dispelled by attending a single meeting of any local deliberative body, anywhere in North America, on the subject of, for example, traffic and parking in the community. Attend a community association meeting about a new mid-rise condo that might increase local traffic. Go to a planning meeting at City Hall (any city hall) on the subject of locating bike lanes on a busy commuter thoroughfare. Check out a municipal candidates' debate and ask about raising parking fees. Or lowering parking fees. Or keeping parking fees right where they are. I can assure you, the discussion will not proceed steadily to an objective consideration of the relative climate benefits of the various options on the table. In fact, I can assure you that the discussion will not proceed steadily at all.

Here is University of Toronto philosophy professor Joseph Heath, in his review of Naomi Klein's *This Changes Everything*, on the underlying argument propelling this faith: "When it comes to understanding the major problems of our age, one of Klein's central convictions is that *the people are innocent*. As far as my own view is concerned, I guess I'm more inclined to think that *the people are guilty*. That sounds a bit harsh, so maybe I should say something more like *the people are not innocent*."

What Heath means is that the balance of evidence doesn't support the hypothesis that people would act in the best interests of themselves and the planet if only they were liberated from the shackles of corporate greed and political corruption. People mostly tend—"ever so slightly," as Heath puts it—to be selfish. They care a little more about their own interests than the community's or the nation's or the planet's. Such ever-so-slightly selfish actions, taken together, create what political theorists like Heath call "collective action problems." And climate change might be the single biggest, thorniest example of such a phenomenon in history. (I'll come back to this later.)

People acting in what they perceived as their best interests made the light-duty truck and the 2,500-square-foot split-level suburban home the default transport and housing modes across North America. People oppose bike lanes and pack community association meetings to discourage adding density to their neighbourhoods.

They do this as readily and forcefully in Berkeley, California, and greenest Vancouver as in oil-rich Houston or Calgary. People, as Heath himself points out, have been known to protest vehemently against wind farms. (In Ontario, public discussion about renewable energy grew so toxic that, in at least one rural township I passed through a few years back, they even protested against solar panels.) People are not rational actors, they are not particularly skilled at assessing their best interests, and they are exceedingly difficult to round up into anything like a lasting consensus, especially on an issue as maddeningly multi-faceted as climate change.

There is a sort of royal "we" that is quite common in the climate-solutions discourse. You hear it in lectures like the one David Wallace-Wells delivered at Globe, in the pages of books such as Klein's. In op-eds and Twitter posts, climate scientists and advocates speak often about what *we* need to do or what *we* are demanding of the world's political leaders or how *we* must reduce emissions. This is a common rhetorical shortcut, of course, but its prevalence in the climate debate verges on real obfuscation. The royal climate "we," writes British science journalist Oliver Morton in *The Planet Remade*, "supposes that my interests are your interests, and that the interests of people in different countries and with different views . . . can be easily aligned. The 'we' . . . seeks to speak for the world." In essence, the royal climate "we" recurs so often because it serves as another way around the intractable problem of political will. Rather than get into that morass, let's talk instead about what *we* would be doing if not for *them*. And never mind that, political will is everything.

Consider the rallying cry of a new generation of climate activists: "Listen to the science." Sounds easy, doesn't it? Perfectly reasonable, eminently rational. The science has been clear since NASA climate scientist James Hansen first presented it to the US

Congress in 1988. It was clearer still at the launch of the Rio Summit—the first global climate conference—in 1992, at the Kyoto conference in 1997, at the Copenhagen climate talks where the climate treaty nearly collapsed for good in 2009. Climate scientist Katharine Hayhoe, who runs the Climate Science Center at Texas Tech University, sums up the science in just ten words, borrowing a line from her colleague Jean-Pascal van Ypersele, vice-chair of one of the IPCC's working groups. Here are those ten words: "It's real. It's us. Experts agree. It's bad. There's hope." I've seen Dr. Hayhoe present this clear, concise summary to several audiences, and no one disagrees. But here's something else she always tells her audiences: the science isn't enough.

"I am a climate scientist who has spent a lot of time trying to make climate science more accessible," Hayhoe writes in a 2018 opinion piece in *Science* titled "When Facts Are Not Enough."

> I've authored National Climate Assessments and numerous outreach reports; I host a YouTube show called Global Weirding; I tweet; I've even promoted knitting patterns that display rising temperatures. Yet the most important step I've taken to make my science communication more effective has nothing to do with the science. As uncomfortable as this is for a scientist in today's world, the most effective thing I've done is to let people know that I am a Christian. Why? Because it's essential to connect the impacts of a changing climate directly to what's already meaningful in one's life, and for many people, faith is central to who they are.

What Hayhoe is saying is that most people don't rationally process information about a heavily charged issue like climate change. They don't evaluate data, analyze its implications, determine the actions that best serve their interests and act accordingly. They address the topic *socially*. They look to see how people who share their worldview—fellow Christians, for example—are responding. And then they act (or don't) in lockstep with their social group. And that too is science.

"This social conformity is not some preference or choice," George Marshall writes in *Don't Even Think About It*.

> This is a strong behavioural instinct that is built into our core psychology, and most of the time we are not even aware that it is operating. It originated as a defense mechanism during our evolutionary development, when our survival depended entirely on the protection and security of our social group. Under such conditions, being out of sync with the people around us carried a potentially life-threatening danger of ostracism or abandonment.

Many of us likely think our own beliefs aren't so irrational, so dependent on conforming to the points of view of our peers. How did I come to pursue climate solutions, after all, if not by evaluating the available information and deciding on the role I could play in adding to it? Surely I was at the mercy of nothing more than my own reason? Perhaps. But reason itself is, as Elizabeth Kolbert notes in a 2017 *New Yorker* article on the latest scientific research examining why we ignore facts, "an evolved trait, like bipedalism or three-color vision. It emerged on the savannas of Africa, and has to be understood in that context."

Here is Kolbert summarizing the findings at the core of *The Enigma of Reason*, a 2017 book by cognitive scientists Hugo Mercier and Dan Sperber:

Humans' biggest advantage over other species is our ability to cooperate. Cooperation is difficult to establish and almost as difficult to sustain. For any individual, freeloading is always the best course of action. Reason developed not to enable us to solve abstract, logical problems or even to help us draw conclusions from unfamiliar data; rather, it developed to resolve the problems posed by living in collaborative groups.

It evolved, as Mercier and Sperber put it, as an adaptation beneficial in the "hypersocial niche" in which people—all people, even scientists themselves—exist.

A library's worth of climate science and a thousand UN conferences won't amount to much without cracking this aspect of the problem. The political will so often missing from the climate equation is a subset of this hypersocial reasoning, a commodity produced entirely from finding ways to turn these behavioural tendencies from liabilities into assets. And there are proven ways to produce this commodity. There is science well worth listening to on the subject.

2.15 Hearts and Minds

When George Marshall asked Daniel Kahneman, one of the world's foremost experts on the psychology of decision-making, what he thought of the prospects for a comprehensive global response to the climate crisis on the timeline and scale required, Kahneman's response was blunt. "I am deeply pessimistic," he said. "I really see no path to success on climate change."

Kahneman won a Nobel Prize in economics with his research partner, Amos Tversky, for their ground-breaking work on the application of "psychological research into economic science, especially concerning human judgment and decision-making under uncertainty," as the Nobel citation phrased it. Which is to say there are few scientists who understand as much as Kahneman does about how human populations collectively choose to respond to what they now know about climate change.

In the years since I first ventured into the political arena around 2008, there's no single source on how change happens that I've learned more from than Kahneman. Since he and Tversky (who died in 1996) received their Nobel laurels in 2002, their work has given rise to an entire academic niche known as behavioural economics, which resides at the intersection of behavioural psychology and economic theory. Their work provides vital insights for any effort to encourage people in large numbers to lend support (and votes and money and enthusiasm and everything else) to new and risky-looking endeavours, especially those aimed at solving a problem like the climate crisis, which only a fairly small slice of the public understands well even now.

I'll try to explain the central concept of the work that led to Kahneman and Tversky's Nobel Prize—a concept they called "prospect theory"—as clearly and simply as possible, because so many of the lessons that behavioural economics has to teach us about how to solve the climate crisis hinge on it.

Historically, as I said, economic theory had been predicated on the presumed existence of a population of more or less rational actors, and those rational actors were presumed to value a dollar more or less the same as any other dollar. If the reality of human behaviour was slightly more complicated, you could still base some very elaborate, effective models of their economic activity on those assumptions. Kahneman and Tversky's prospect theory showed instead that the value of a dollar is determined situationally, based on its reference point, its context. It matters whether the actor involved has recently lost or gained wealth. And even more important, it matters whether the dollar in the equation itself is perceived as a potential loss or gain.

The feature of prospect theory most important to the climate discussion is "loss aversion." Here is Kahneman's definition of this principle: "When directly compared or weighted against each other, losses loom larger than gains. This asymmetry between the power of positive and negative expectations or experiences has an evolutionary history. Organisms that treat threats as more urgent than opportunities have a better chance to survive and reproduce." The principle sounds simple enough—"losses loom larger than gains"—and easy enough to incorporate into an organization's decision-making or a government's public messaging. But when this loss aversion combines with many other habits of mind and our evolutionarily developed use of reason as a tool for social cohesion, the result is a tangle of mental habits, shortcuts and biases that make climate action an exceptionally hard sell.

This might sound counterintuitive. Isn't climate change a threat of the deadliest gravity? Doesn't it promise to create catastrophically large losses, not just of a healthy and stable natural world but also property, wealth, human life? Absolutely. But the climate threat is not *perceived* as such, at least not widely, and so the actions proposed to prevent it are seen as a potential loss. Climate change possesses none of the four crucial qualities that psychologist Daniel Gilbert has identified as the primary triggers of humanity's threat response: It is not a direct personal threat. It is not abrupt. It is not, in most people's estimations, immoral. And it's not happening right now.

In the absence of Gilbert's four horsemen of the psychological apocalypse, the climate threat is instead processed by brains designed by evolution, cognitive processes and deep habit to mis-apprehend it and prefer the status quo. The human mind, riddled with a wide range of biases, shortcuts and simplifying habits (which Kahneman and his colleagues have catalogued exhaustively), is strongly predisposed to see the crisis as far less threatening than the potential losses caused by responding to it in any significant way.

Among the most important—and most common—of these habits of mind detailed in Kahneman's *Thinking, Fast and Slow* is the "anchoring effect," which occurs, he explains, "when people consider a particular value for an unknown quantity before esti-mating that quantity." The asking price for a house, for example, "anchors" its value in your mind before you've ever considered its attributes or the values of similar properties. In an experiment that behavioural economist Dan Ariely has conducted many times over with his classes at Duke University, students are asked to write down the two digits at the end of their Social Security number—a wholly arbitrary value—and then consider the bidding prices for a selection of luxury goods of indeterminate value, such as French wines and fancy chocolates. The students whose final two Social Security digits are higher will invariably bid higher for the items on offer, simply because they've anchored their minds to higher num-bers. The brain is just that easily tricked.

Beyond this, there are numerous other common biases, from the "judgment heuristic" (a mental shortcut by which people will routinely pass judgment on, for example, the competence of politi-cal candidates based entirely on the shape of their chin or the set of their mouth) to the "halo effect" (a tendency to assume that some-one with one admirable trait is admirable in all respects) to the "sunk-cost fallacy" (referring to the way people make present-tense decisions based on how much time, money or emotion is already invested in a choice). All these biases and mental tics matter, because they affect how we respond to the threat of climate change and assess the viability or desirability of embracing change and adopting solutions. The assumption that appeals for climate action are reaching people who are ready to evaluate scientific evidence accurately, understand the data clearly, and respond rationally in their best interest is as fanciful as the rational actor of outmoded economic theory. *No one* just listens to the science. And the science itself—psychology, I mean, not climate science—is unequivocal regarding why.

Of particular note for the climate-solutions beat is understanding how deeply wedded people tend to be to the status quo, to how things currently happen to be. People are by nature inclined not to strive for big change at generational scale, as the global energy transition requires, but rather to want to keep things just the way they are for themselves and their peer group. This "status quo bias" is amplified by what behavioural economist Richard Thaler has called the "endowment effect," referring to our tendency to value the things we possess much more highly than we would if they didn't belong to us. This endowment effect figures enormously in our collective responses to the climate crisis.

Every proposed change—solar panels in place of big power plants, electric motors in place of internal combustion engines, biking or the bus instead of driving, high-efficiency appliances that cost more upfront but earn back the extra cost and more over their lifetimes, and on and on—every one of them triggers another recurrence of the endowment effect. The systems that sustain us now are highly valued. Electricity on demand, a car to reliably take you wherever you need to go whenever you want, a warm home in winter and a cool home in summer, an annual escape to the Caribbean or RV trip to the lake—any one of these would be subject to a strong cognitive bias against change. Cumulatively, they are the greatest prize. They are life itself.

From the point of view of cognitive instead of climate science, there is no mystery as to why people aren't rushing into the unfamiliar void of the single greatest change humanity has ever been asked to make. Add the slightest sheen of uncertainty, the barest tinge of misinformation or negative feedback, and the mental barriers become all but impenetrable. The first "green" cleaner you tried wasn't as effective as the old reliable. You saw a news clip on Facebook about an electric car bursting into flames. The person you most closely associate with solar panels is a loathed politician or an eccentric counterculture type (or even—*yikes*— a European).

Compound the endowment effect with judgment heuristics and sunk-cost fallacies and our general tendency not to be very good at assessing long-term risk and comprehending stats and scientific arguments, and *of course* climate solutions are among the very hardest of sells. The wonder, really, is that change is already under way at such a pace. Two houses on my block in the heart of Canada's oil patch have solar panels on their roofs. I see Teslas in Calgary all the time now (one with the vanity plate LOL OIL). Canada is well on its way to eliminating coal entirely from electricity production, and there's a price on carbon nationwide. In the face of a towering cognitive wall of bias and habit and the myriad status quo loyalties of party and region and tribe, somehow the new and unfamiliar and uncertain have busted through. Look on in awe. It means that a lot of the hard work has been done already.

There is probably no package of policy shifts, consumer changes and social reorganization that challenges people's basic assumptions and threatens their livelihoods and sense of social position and self-worth as strongly and with such sustained pressure as climate change does. Climate change makes villains out of heroes, turns producing that most vital of daily needs—energy—into an evil enterprise. It pits parents and grandparents against their children and grandchildren. It deepens inequalities and expands existing divisions between rich and poor, industrialized and developing, north and south. And it turns region against region. (I live in a province where gas and coal still fire the electricity grid and oil production drives the economy. I could move a three-hour drive west and flip instantly from villain to saint in hydro-powered, cleantech-focused British Columbia.)

Climate change calls into question every single aspect of daily life, from the flow of your showerhead to the soap you lather up with to the type of refrigerator you pull the orange juice from (cardboard or plastic? organic or not?) to the coffee beans you grind to the way you commute to the type of office building you arrive at to the kind of business you conduct there. Of course climate solutions are hard sells. No one wants everything they do every damn day put under that kind of scrutiny, and no one wants to spend that much of their day thinking about process. They just want to get on with it. The shift from the less bad approach to the much better world is in large measure about getting on with it, making the climate-friendly choices uncomplicated and ubiquitous and *automatic*.

And as for the political will to make that happen, here's Kahneman on that: "Because adherence to standard operating procedures is difficult to second-guess, decision makers who expect to have their decisions scrutinized with hindsight are driven to bureaucratic solutions—and to an extreme reluctance to take risks." An *extreme* reluctance to take risks—that is perhaps the primary reason why the political will to act faster and more radically to address the climate crisis has been so lacking in most of the world for more than twenty years.

Upending bureaucratic systems entirely to accelerate a government into emergency mode would come with astronomical political costs. The timeline and scope of the climate crisis guarantee that the full benefits of such decisions would not be apparent anywhere near in time for voters to properly reward leaders who took such action. On the contrary, the much more likely outcome would be a reactionary opposition's rise to power on a vow to undo the radical changes, as has happened in numerous jurisdictions around the world. (Perhaps the strongest case in point is Australia, whose ahead-of-the-curve carbon tax policy was immediately torn to shreds by the government that toppled the one that put it in place.)

Small wonder, then, that Daniel Kahneman is a pessimist regarding our collective ability to tackle the climate crisis. I'd be pessimistic too if I thought the only way to address it was to sell the entire world on punitive climate policies in a hurry, to win support to act argument by exhausting argument, to dismantle the Great Wall of Biases brick by mental brick. But there are other ways, other incentives that line up those biases in favour of action. That's where I find my optimism. More of it by the day.

3.0
THE MUCH BETTER
DECADE

3.1 The View from Quarantine

I had plans for 2020. We all had plans.

In the last few months of 2019 I'd been combing through library databases and the less obvious results of my Google searches, searching for an uncommon vista from which to properly assess the state of the climate-solutions beat and the prospects for the decade ahead. I wanted the best possible view of a rapidly brightening horizon.

I stumbled on an absurd place, a fantastical slice of tropical paradise. It was an item on the travel blog at *Forbes*, of all places, basically a real estate listing for a $13-million home. On an island in Hawaii. Gated community with its own golf course. Five thousand square feet on more than half an acre. Infinity pool, home theatre, "chef's kitchen." Ridiculous.

But the headline read "Tesla Owners Will Love This Stunning Home on Kauai," and I had to admit I loved it too. I'd seen an idealized artist's depiction of a Tesla-powered future in the company's own marketing materials—modern home, solar-panelled roof, garage full of Tesla Powerwall batteries, all-electric Tesla sedan in the driveway—but here it was in the wild. "An expansive garage that is pre-wired with four Tesla Powerwalls," the *Forbes* post gushed, as if they were just another standard luxury amenity to go with the infinity pool.

I quickly learned that the island of Kauai was well on its way to becoming the first renewable-energy-powered island in the state with America's most ambitious renewable energy plan. Kauai was leapfrogging from energy laggard to global leader. I was hooked by the juxtaposition of postcard-flawless natural scenery and preening luxury, the future-is-now Tesla garage on a remote island long troubled by high power prices and uncertain energy supplies. This felt like the start of something, not an outlier but a leading indicator. I decided I would make my way to Kauai (tough gig, I know) and survey the state of the climate-solutions beat from a Tesla parked alongside a lushly landscaped lawn, overlooking the infinite possibilities of an overflowing pool whose lip merged seamlessly with the horizon itself.

And then, of course, there came a pandemic, and that was the end of that. So I tracked the progress of the global energy transition—this object in rapid motion—from a screen in my home office, and after a year and a half (and counting) under the dark shadow of the coronavirus, I'm happy to report it's still a powerfully inspiring view. The pandemic only strengthened my sense that a much better world is on its way. Not because the race to develop a vaccine was so inspiring, and not because the reluctant, uneven efforts to control the virus's deadly spread were sometimes so heartening in their communal spirit—though both are true, and I think in retrospect the extraordinary collective effort billions of us all made to stop COVID-19 from taking a toll that could have been many orders of magnitude more brutal will be a source of some inspiration. No, I'm more optimistic than ever because the mood yawning out from under the pandemic's dark weight has been so ripe with possibility.

"Build back better." That was one of the first rallying cries that united government-led efforts around the world to plan for a rapid rebuilding of the post-pandemic economy. The phrase originated in a UN report prepared to plan for reconstruction after the 2004 tsunami, but it came into widespread use in the summer of 2020 as COVID case counts declined and viable vaccines emerged. It pops up in press quotes from British prime minister Boris Johnson in May 2020, and Singapore's Reform Party employed it as an official campaign slogan in June. And then Joe Biden began using "Build Back Better" as a tagline for his presidential campaign—it fronted the lectern for the cameras, as political slogans now do, at a rally in Rust Belt Pennsylvania on July 9 that laid out his economic plan—and it has since expanded, meme-like, into widespread use.

The European Union's pandemic recovery plan has dedicated 30 percent of its funding—more than €500 billion over seven years—to "climate-facing" initiatives. Weeks after Joe Biden became president, his office unveiled an infrastructure and pandemic aid package worth more than $3 trillion, with climate solutions central to the new funding. "Biden's Recovery Plan Bets Big on Clean Energy," read a *New York Times* headline. Earlier, in the depths of the pandemic, South Korea and New Zealand elected governments that explicitly promised to harness their economic futures to climate solutions.

In the fall of 2020, an IEA report noted that the only part of the global energy industry that continued to grow during the pandemic was the renewable energy sector. Around the same time, NextEra Energy, a developer of large-scale renewable energy projects, was briefly America's largest energy company by market value, beating out ExxonMobil, which had held the spot for decades. One car company after another, meanwhile, had taken to trumpeting their commitment to expand electric vehicle manufacturing. Hyundai promised to sell 500,000 EVs per year by 2025, and Volkswagen vowed to produce one million EVs per year starting in 2023. It culminated in an announcement by General Motors, timed to coincide with President Biden's ambitious postpandemic climate plan, that by 2035 it would be manufacturing electric vehicles exclusively. The cost of battery storage, currently around $140 per kilowatt-hour on average, is plunging, and within a few years it should fall below $100 per kilowatt-hour, a crucial threshold estimated by the industry to be the point of no return, guaranteeing that the clean energy world will soon have as much affordable large-scale storage as it needs.

A loose consensus has emerged, and it began in the wake of the Paris climate summit in 2015—the nearest the world's political and business elites had ever come to universal agreement on the necessity for climate action. At the time, that consensus looked exciting, if precarious. But it has held. It held through the frothing absurdities of Donald Trump, who ran for president on making "clean, beautiful coal" great again and dismantled American environmental regulations to the precise specifications of the oil-and-gas business. It survived other blustering authoritarians, from Russia's Vladimir Putin to Brazil's Jair Bolsonaro. It endured Brexit and Boris Johnson (the latter, notwithstanding his many faults, is a long-time urban cyclist who seems fine with the rapid decarbonization of Britain's electricity grids). The line held, to use a war metaphor. More than that—it got stronger.

And though it did so for a great many reasons, most of all it held because renewable power was cheap and getting cheaper, because electric cars were becoming affordable and growing more popular, because big business likes efficiency improvements as much as pioneering green architects do. The global energy transition is well under way, and it is unstoppable, and it has proven that it can readily meet the needs of a technologically advanced, power-hungry world of nearly eight billion people and growing. Even the oil-and-gas business is, in its internal discussions, increasingly talking about when, not if, the pivot will occur. The consensus held because it was the right thing to do, and the smart thing to do.

The decade to come will witness the first full global push of the fastest and most comprehensive—and potentially most inspiring—transformation in the way the world produces and uses energy in all of human history. This was not only far, far, *far* from inevitable (far!) but seemed quite improbable when I first surveyed the newborn transition nearly twenty years back. There were two massive obstacles barring the way forward, and both have been overcome. The first was direction—the level of loose cooperation among the world's nations and industries necessary to turn the whole planet's emissions trajectory downward. The second was speed—which, as slow as the energy transition might seem from the front lines of a Fridays for Future march or the chart of any given nation's annual emissions, is actually hitting a pace that will far exceed any previous energy transition since we first captured fire. A pace more than fast enough, I'd say, to fuel a whole world's optimism. And a direction that I find very exciting indeed.

3.2 Reality Check

Let's talk about what *excitement* means in the context of the global energy transition. I don't find electric car manufacturing and energy efficiency retrofits and corporate net zero pledges to be intrinsically exciting. I'm speaking relatively. I spent twenty years with a close-up view of the status quo's inertia, and *that* was not at all exciting. Being told that solar power was a treehugger's luxury and electric cars a distant-future fantasy and transit too expensive and bike lanes for Europeans and hobbyists, all while governments were rubber-stamping oil sands mines and fracking operations and highway expansions and outer-belt suburban development—none of that was exciting. That the global energy transition is building momentum to the point that not even oil and coal companies can ignore it anymore is exciting in that context only. But it is—in *that* context—very exciting.

As I've said, it was all hard to imagine when the inertial forces weighing down change seemed at their heaviest in the past ten years. It seemed, at times, that the only option was a sort of wishful thinking—an accurate read of the scale of the problem and a somewhat fantastical notion of how it might be overcome, all at once. A sort of utopian lens. I think, for example, of Paul Hawken, co-author (along with Hunter and Amory Lovins) of *Natural Capitalism*, published in 1999, the first book I ever read on climate solutions. Hawken has been called "one of the leading philosophers of the sustainability movement," and his thinking has inspired any number of people and organizations developing the toolkit for a much better world. He inspired me.

Hawken's current project is called Drawdown. His 2017 book of the same name is subtitled *The Most Comprehensive Plan Ever Proposed to Reverse Global Warming*. It might well be. Hawken and his Project Drawdown colleagues have done deep, comprehensive work on the paths to lower greenhouse gas emissions. Their data is impeccable and regularly updated. Surprisingly, "refrigerant management" landed at the top of their list of ways to cut emissions; discarded refrigerants are a significant source of greenhouse gases in landfills. The other solutions clustered near the top of Drawdown's list include better-known stuff—solar and wind power, net zero buildings, urban design tools like walkability and transit. And yet there remains, even in Hawken's work, that maddening gap between what "needs" to be done and the political, economic and social levers to make it happen.

"This comes back to language," Hawken told a Vancouver "eco-city" conference in 2019. "How we're talking to each other about climate change I feel like has disengaged 98 percent of the people in the world. . . . How did we do that? How did we create a movement where we are addressing the greatest crisis civilization has ever faced and probably will—and have that much of the world disengaged?" I waited to hear about political barriers, about behavioural psychology, about the systemic biases and peer group loyalties that have slowed climate solutions for a quarter century.

"We're still not naming the goal," Hawken said.

> We're using verbs. *Mitigate, fight, tackle, combat.* Those are not goals. . . . There is no battle and there's no one losing, so there's not a fight. Because as soon as you use that language, basically you are dividing the world. You're using male war and sports metaphors to talk about something that's exquisite and beautiful, which is the climate. What a gift it is. And climate change is not the problem. Human change is the problem.

Change, he argued, will come from "a community of people together, talking, everywhere in the world. And knowing that they're the ones who are going to make a difference."

I want Hawken to be right, for all the dreamers chasing not just a more stable climate but a fairer, more just and more equitable world to be right. But I can't come around to seeing the solutions to a collective action problem about energy use and pollution fixing all that along the way, not on a timeline of a couple of decades. I find myself drawn instead to a conversation I had with Joseph Heath about collective action problems and what he calls "Hobbes's difficult idea."

That's Hobbes as in Thomas Hobbes, the English Enlightenment philosopher best known for his 1651 treatise *Leviathan*, in which he argued that humanity's natural state is one of warring anarchy. "Hobbes is so important in this," Heath told me.

> Because what he corrects is a sort of common misunderstanding people have, which is we assume that people are automatically cooperative, so that if it's in our common interest to accomplish something, that everybody will just sort of automatically do their part and we will have this collective achievement. But all of the misery in human history is just a consistent litany of people failing to do that. So that's the basic collective action problem idea, which is that there are a lot of circumstances in which it's our common interest not to do something but it's in our individual interest to do it. That's the free-rider incentive that leads to failures of cooperation.

The free-rider problem is a subset of what political scientists call collective action problems. The best-known example of these is the "tragedy of the commons," a reference to the way farmers in medieval England would allow their sheep to graze the town common to ruin because, although they all had a collective incentive to keep

the common green, none had an individual incentive to stop let-
ting their flock chow down on the common's lush grasses—not if
all the other shepherds were carrying on regardless. Overfishing is
an excellent modern example of an environmental collective action
problem. Every nation's fishing fleet needs abundant fish to sustain
itself, but in the absence of universal catch limits and fish-stock
management, individual nations and individual boats are chasing
a quarry that swims across jurisdictional lines. No single country
or its fishers have an incentive to reduce their catch if the rest
intend to continue with business as usual. And so, in the absence
of incentives to cooperate, the world's fish stocks are depleted.

Heath again on the difficulty of such cooperation at global scale:

Humans have only gotten good at cooperating with each
other in the last 200 years. The market is a system of cooperation, the welfare state is a system of cooperation. For most
of human history, the reason people were poor is not because
they didn't have stuff. It's because they didn't cooperate with
each other. The absence of cooperation is the baseline, and
the accomplishment is when we manage to cooperate with
each other. So climate change as a collective action problem—
the default setting is that nobody does anything about it
and we wreck our ecological niche. The achievement would
be if we manage to cooperate with each other in order to
control the problem.

To my mind, Heath is not being pessimistic here. He's saying it's
extraordinary—a staggering achievement, really—that there even
exists a sufficient level of cooperation, through international markets and institutions like the UN, to coordinate a weak, messy, eter-
nally hamstrung global response and share useful tools. But the

essence of Hobbes's difficult idea and the repeated lesson of history is that there will be division and competition—especially at global scale. And so, wishing it were otherwise, I look for excitement on the climate-solutions beat, wondering less about how to inspire people to change and more about how to incentivize them to change. How to sell them on this energy transition, I mean. Sell it like Coca-Cola or a smartphone or a car racing down a winding open highway.

3.3 Pledge Drives

When I first started on the climate-solutions beat, many of the innovations I discovered—renewable energy, more efficient buildings, dense and walkable urban design plans—were branded as *green*. Green energy, green buildings, green cities. It's a label that persists, though its altruistic tinge and fringe appeal have faded considerably. Green stuff back then was still mostly for green people—environmentalists and pinkos, vegetarians and hippies and off-grid zealots. Your nearest "natural food" store probably still has a few aisles' worth of this stuff, bags of pasta and tubes of lip balm with watercolour scenes of natural abundance and a brand name that tells you how good it is for the planet or what a strange and wonderful character its original maker was. (The Burt's Bees line of skincare products has been a subsidiary of Clorox since 2007, but there's still a pen-and-ink portrait of a bearded hippie on the tin.)

As green things moved to the mainstream and seeped into corporate offices, *sustainability* became the preferred term for doing less bad on the environmental front and beginning to take at least some interest in the climate costs of doing business. There were many impressive achievements under the sustainability banner: Patagonia made mountains of fleece from old pop bottles, Walmart overhauled its entire trucking fleet for maximum fuel efficiency, companies by the hundreds commissioned green headquarters, and the United Nations unveiled a long list of ambitious sustainable development goals, to name a random few. But as a concept it was fuzzy enough that any company or organization could plant some trees or donate to a conservation group and then claim to have embraced sustainability.

Numerous other conceptual frameworks were developed to somehow codify the range of complex processes by which a business or organization might reconcile itself with the enormity of the climate crisis. There were (and are) *triple bottom lines, resilience plans, carbon footprint measurements, corporate social responsibility departments, B Corp certifications*. The most recent boardroom buzz-phrase to gain wide sanction is *ESG*, short for *environmental, social and corporate governance*. And under each of these banners (sometimes seemingly all of them), sincere efforts were initiated and often real progress was made in making a company more energy efficient or shrinking waste or protecting wetlands or commissioning wind farms.

It's easy to be cynical. But as my old friend Alex Steffen, a climate futurist who's been thinking about this crisis as long as anyone I know, likes to say, "Cynicism is obedience." Cynicism reinforces the established order and shrugs off challenges to the status quo. All those first-wave efforts at some kind of climate action, however modest or even deeply cynical, did accomplish something very important in the aggregate. They initiated uncomfortable conversations in boardrooms and legislative offices. They enabled groups of people who were unlikely to embrace big green ideas right away to get used to living and working with much smaller ones. They permitted trial and error. They laid down preliminary plans. Like a network of small, seemingly unconnected fissures on the face of a towering dam, they prepared the vast structure to allow a much more transformative force to burst through.

All of this is my slightly meandering way of coming around to my reaction in April 2020 to a news item about Royal Dutch Shell. The world's fourth-largest oil company was committing to a net zero goal for its greenhouse gas emissions by 2050 at the latest. I would submit that there are few announcements to which cynicism can be as effortlessly directed as an announcement by a massive oil company that it will one day generate no emissions. The source obviously invites cynicism. The timeline, thirty years off—distant enough to be science fiction while still falling within the general scope of current climate action plans—invites cynicism. And the concept itself—the *net* part of it in particular, a target hedged against the expectation that future-tense technologies or some colossal tree-planting program will be able to offset the substantial emissions the company might still be generating in 2050— invites cynicism.

And yet my reaction was closer to *Hell yeah!*

I should emphasize just how unlikely it was, even ten years ago, that a top-ten global oil company would promise that by mid-century the entirety of its operations would no longer generate carbon dioxide. How completely outside any kind of discussion within the industry such a goal was, let alone that it would be embraced, formulated, publicized, slapped up on the old website beneath the logo with a bunch of targets and plans and such. You take my point here? This is about the pace of change, and by that measurement it's lightning quick.

Royal Dutch Shell was far from the first major multinational corporation to announce a net zero goal. It wasn't even the first big oil company to make such a pledge—BP had done so in February. Google, Apple, Amazon and Microsoft have net zero pledges (Microsoft is actually promising to go beyond net zero, to compensate for emissions all the way back to its founding in 1975). A great many governments around the world—Denmark, the United Kingdom, Japan, South Korea, New Zealand, Canada—have their own "net zero by 2050" pledges (Sweden and Germany cheekily set their targets for 2045). And they are all joining a vast range of solemn pledges and bold targets: 50, 80 or 100 percent renewable energy on the grid; electric vehicles alone available for sale; every new building constructed to net zero standards. Even China has promised to reach net zero, though it isn't planning to hit the mark until 2060. Twenty-five years off, or perhaps thirty or forty— it's all sufficiently far in some fictitious future to allow plenty of time to further hedge, qualify, obfuscate, renege. How could you not be at least a little cynical about it? "A credible net zero plan from Shell would start with a commitment to stop drilling for new oil and gas," a British Greenpeace spokesperson told Reuters when the company's target was first announced. You could pretty much copy and paste the same sentiment into a wire report about any of those pledges. And yet, as I said, I heard the news, saw this great wave of net zero promises and thought: *Yes. Here we go. This is the point of no return.*

I do not mean to imply that Shell and Microsoft have come to save us all. I simply want to underscore how significant it is that the architects and primary benefactors of the fossil-fuelled status quo are now signing pledges for its termination. The reluctant, multiply hedged, not-necessarily-near-term promise of an oil company today isn't the same as a feel-good tree-planting project in 2005. I haven't picked that date arbitrarily: My first child was born that year, and at the time BP was sponsoring tree plantings all across Calgary, dedicating one to each newborn in the city. We attended the ceremony. The Calgary Police Service parked its ceremonial Hummer there for kids to gawk at while we planted trees far too insufficient in number to make up for even just the damage wrought by the hulking SUV's big tailpipe. It was a different time. But it was, you know, nice enough—a baby-sized first step on behalf of our baby.

How do I mean it's not the same? I mean that there was no target for those tree plantings, no final measure of success. An arbitrary number of trees theoretically compensated for an arbitrary volume of carbon dioxide emissions, and that was enough to add a colourfully illustrated blurb to the annual report. Net zero is a hard target. Shell can miss it. Denmark and China and Canada can miss it. The world can miss it. But it's clear, meaningful and relatively easily measured. It's a much better pledge. And it's a tacit admission by fossil fuel companies, automakers, manufacturers and sellers of consumer goods, heretofore largely indifferent to their climate impacts, that they will not be in the same business—at least not in the same way—within a quarter century or so.

Listen. Net zero has real teeth. I wasn't completely clear on this myself until I saw the process up close. Not long after Canada made its net zero pledge—a blithe and easily derided campaign promise as the federal Liberals ran for re-election in 2019—an organization called the Canadian Institute for Climate Choices embarked on an economy-wide examination of the plausibility of the goal. (Does the think tank's name sound more than a little fuzzy and unaggressive and inconsequential, particularly in comparison to, say, Extinction Rebellion? Good. It's supposed to.) When the Institute was done with its modelling, data-crunching and analysis, I was invited to assist with producing a final report that would translate the wonkery into a story an interested layperson could understand. And let me tell you what the report said, beyond its many data sets and graphs. It said that Canada's net zero goal was achievable, mostly with existing technology, and that it would likely be of net benefit to the Canadian economy—in addition, of course, to being an existential necessity.

What was most powerful about that report, though—about working on it, I mean—is that it was like having a team of hard-nosed science teachers check my work from fifteen years on the climate-solutions beat and award me a passing grade. The solutions *were* mostly there. They *did* point in a better direction, from the street to the boardroom to the legislature to the overheated skies above. And there was real bite to it all.

This wasn't a catalogue of admirable one-off renewable energy projects on remote islands or wishful projections about electric car ownership. The modelling behind the report looked at how cement and steel are made in a net zero Canada, how farmers tend their crops, how freight is transported from one end of the vast country to the other, how cities grow more efficient and provide abundant employment, how homes stay lit and warm in the cold Canadian winter. Easy as it might be to remain cynical about it all, this was a blueprint for a much better world. Which is, yes, pretty damn exciting work to be a part of.

3.4 Objects in Motion

I spent my high school years in North Bay, a small city in northern Ontario. The winters were long and mean. I lived in a new subdivision directly behind the city's newest high school—there were still open-pit foundations for new homes all around us when we arrived in the summer of 1988—but I attended a Catholic school far across town, and my friends were scattered all over the city. My bus stop was right in front of that newer high school, and I often had time to contemplate the frustrations of suburban distance as I waited in the biting cold for the infrequent city bus that would transport me to wherever my friends were hanging out that night. On weekend nights the frequency was one bus per hour; miss it by a minute as you tried to avoid unnecessary time out in that frigid air, and you were an hour late for the evening's fun. There was usually a transfer required at the main transit depot downtown, which often meant another five or ten minutes with that singular sensation of numbness turning to an ache in your inner thighs as the wind chapped them red through two layers of clothing.

I turned sixteen the following summer, and I took my driver's test within days of my birthday. In North Bay in the summer of 1989, a driver's licence was freedom, plain and simple. My friends and I needed no marketing campaign to sell us on the concept, no coercive lobbying by oil companies to make us not just willing but eager—desperate—to pull up at the local Shell station and hand over a full evening's wages from a part-time job to fill the tanks of our parents' cars. The licence and the full tank meant a Friday night of driving up and down the main drag, stopping in at the McDonald's parking lot on Lakeshore Drive, where teenagers gathered if there was nothing else to do. It meant pointless weekday evenings driving out into the country, bombing forty-five minutes down the highway to the next town to buy snacks because it was something to do, driving to each other's houses to watch movies, driving to pick up a special someone to do little more than go driving and park somewhere quiet.

To put this in behavioural economic terms, all the heuristics and biases fed a simple equation: car = freedom. You could've given away bus passes, even made them redeemable for free fries at the McDonald's, and still no teenager in North Bay would've been willing to take that deal over an evening of access to Dad's K-car. This is the kind of deep-in-the-bones inertial force—economic and physical, cultural and social—that an energy transition on the scale now under way must inevitably confront. The expense has already been built into the existing way of life by design and by habit, through countless cultural cues, and in an anchoring effect that clearly indicates the price of a tank of gas is more than reasonable for all it delivers.

The status quo bias at work here is simply massive. It's not just about the relative utility of cars in small cities with poor transit infrastructure. It's about what people trust, who they believe themselves to be, how they think a daily life and a lifetime ought to be shaped. The switch turns on the lights, the thermostat turns up the heat, and trusting anything other than what is currently delivering that comfort and safety is a very big change. Houses that stand by themselves (ideally with a decorative expanse of freakishly uniform grass) are preferable to dwellings attached in a row, which are preferable to ones that are stacked in a tower. Cars have engines and gas tanks. Buses are infrequent and uncomfortable. A street is for vehicles, cars and trucks, commuters and shipping. To run a successful retail business, you need parking spaces. An office park demands an expansive parking lot. The list of all the built-in biases, structurally ingrained habits and hard-wired design preferences of the system as it is are pretty much endless.

I don't mean to say they can't be changed. I mean to emphasize that change on this scale, at this pace, to this degree, is extremely goddamn hard. And resistance to that change isn't crazy or deluded, and it's not primarily the fault of a coercive government or a misinformation campaign or the false consciousness engendered by the "100 companies responsible for more than 70 percent of the world's emissions." So yes, let's discuss how the world will change faster than ever in the next few years, how there are cleaner tools at hand with which to build a much better world, and how necessary and even inspiring this transition can be—but let's be clear about the difficulty of the task.

For me, freedom isn't a driver's seat anymore. I can't tell you when that flipped forever in my mind, but it has. I own a car, but it means very little to who I am, other than its utility for fetching groceries and taking the kids to baseball. I drove myself to university in a rapidly decaying 1981 Ford Mustang my parents handed down to me. It finally died on the highway between North Bay and Kingston, and the university town and its compact student district eliminated the necessity of a car from my life for awhile. Then I lived in Toronto for nearly a decade, in apartments over stores or in shared downtown houses, and cars were mostly pure hassle.

I rode a bike on urban streets long before it was safe, not because I had some grand sustainable urban-design revolution in mind but because I'd grown up riding bikes on the quiet streets of military bases and saw them as effective transportation. Got clipped once by a rear-view mirror in downtown Toronto, went flying over the handlebars and took it mostly on one shoulder; I'm probably lucky I didn't split my skull wide open. I rode transit a lot, and though a crowded subway car is hard to love, it mostly did the trick. I lived for a year in India, rode trains and took diesel-belching rickshaws on short urban trips, haggling over the price in clunky Hindi. Then I settled down and started a family in Calgary. We chose to sacrifice size and space for proximity, so we lived near downtown and managed as much of our lives as possible by LRT and bicycle and walking. But cars became a higher priority again because, like many North American cities that did their growing after the Second World War, Calgary is much more car dependent than many other places in the world.

But as I said, a car isn't freedom anymore. Freedom is landing at Frankfurt Airport, collecting my bag, finding a kiosk, and buying a ticket on a high-speed Deutsche Bahn ICE train. Freedom is knowing I can get to nearly anywhere I want from the platform right there inside the airport, without needing to turn an ignition key or step on a gas pedal. Freedom is the AVE flying across the Spanish plain at 300 kilometres per hour. Freedom is borrowing an old friend's bike in Copenhagen and using the best urban cycling network on earth to get almost anywhere I want to go.

Freedom was that afternoon in Svaneke—the last day I'd spend like that before the pandemic—sitting on the main square next to a microbrewery, drinking local draft and reading in the sun and pondering the ease of life in the lap of the twenty-first century's best value proposition for emissions-free existence. I can still feel it—the glare of the sun, a slight bite from the Baltic breeze. Later I'd take a bus back to the hotel at twilight and get a little work done. It was perfection, a scene out of a utopian fantasy I'd somehow been able to live for a single afternoon.

That's the better world I want to help build in the wake of this grinding pandemic. But I know I can't take everyone to Svaneke to explain the appeal. So let's talk now about how to make the case for this transition everywhere else.

3.5 The Green Marshall Plan

Here is another story about the powerfully intoxicating lure of freedom. In the spring and summer of 1992 I lived in southwestern

Germany. My father was stationed at Canada's last Cold War military base, and it was my last few months under parental supervision before I started university. I spent my days cutting grass for the base's roads-and-grounds crew, and in the evening my friends and I often hung out at a local pub, playing pool and making the most of Germany's relaxed drinking-age laws.

German reunification was hardly two years in the past, but young East Germans had already begun to come west, seeking jobs and a wider range of opportunities than they'd ever expected as children growing up in a police state. Every now and then a group would turn up at the pub, seeking a glimpse of the long-forbidden West—not just West Germany but actual North Americans, like the ones who played rock 'n' roll and cavorted gleefully in ads for Coca-Cola and Levi's jeans. They were always keen to talk to us, though their fragmented English and our nearly non-existent German meant little could be communicated beyond a vibe of general goodwill. I remember a sense of being in the spotlight when they were around. We were living models of a cultural and social order that had been both the enemy and the envy of theirs. I could almost physically feel their yearning for the kind of casual liberty we took for granted as Canadian military brats, drinking excellent beer that cost next to nothing, chatting up young women in the pub, buying cheap train tickets to Amsterdam for the weekend to buy weed and goof around.

This will sound, more than a quarter century down the line, like I'm overstating their enthusiasm, drawing a caricature. I can assure you I'm not. I've never before or since come so close to witnessing first-hand the visceral thrill of the liberation of a people. I wasn't there at the Berlin Wall as it was breached and then smashed apart, but I shared casual conversations with a few young men who had raced through the new opening to the nearest facsimile they could find of America. (It wasn't all Coke and Levi's and pool at the pub next to the Canadian base. Decades later, in old industrial towns in the former East that had moved into the solar business, I would meet workers whose families had been torn apart by the economic calamity of reunification, their spouses and children toiling in the distant West and coming home far too infrequently. But the vibe back in 1992 was nonetheless exuberant.)

When I think back on that time, I find myself humming the classic Tom Petty song "American Girl." The reason is kind of silly. A couple of years later, in 1994, my father was temporarily reassigned to Vicenza, Italy, and during my Christmas visit we took a weekend trip to Rome. On the long drive back to Vicenza I couldn't find anything on the car radio other than the insipid Euro disco-pop that seemed to be playing 24/7 from every station on the continent. Finally we came in range of the Armed Forces Network broadcasting out of the US military base in Vicenza, and the opening chords of "American Girl" filled the car. And I felt—this seems a little silly, as I said—but I truly felt the crashing wave of the whole great, overstated promise of the American Dream wash over me. For the first time I fully understood how enormous the enticement must have been for some East German kid recently escaped from the polluted, paranoid, bankrupt GDR to try to insert himself into a Levi's ad at a pub in southwestern Germany.

She was an American girl / Raised on promises / She couldn't help thinking that there / Was a little more to life / Somewhere else.

That *little more to life somewhere else*? That powerful yearning? That's the stuff of mass movements, grand migrations, hopefully of global energy transitions accelerating to fast-forward speed. Never mind that the actual destination—the fabled USA, some generic pub on the edge of a Canadian military base, our emissions-free future—is guaranteed to be as flawed and frustrating as any other place built by imperfect human hands. It's that *little more to life somewhere else* that can drive big changes. It's the adhesive that binds together the incentives into a holistic vision of a way of life worth pursuing. People in a panic do not try to reinvent their lives; they try to find safe harbour as fast as they can. People who are invested with yearning for a little more to life can build (or rebuild) a better world.

History gets forgotten very quickly, reduced to a few images and symbols. The full weight of it is hard to recall. Easy to forget that the commercial enticements of the capitalist West looked and felt and tasted like liberation in the heady days of Germany in the early 1990s. Or that—let me rewind further back here—the moral and political authority of America at the end of the Second World War stood as close to uncontested as such status ever can be.

The Second World War can serve as a powerful metaphor for many things. But in climate circles, the wrong Second World War metaphor is in play. There is abundant talk about a war footing. It would be better to talk about a Green Marshall Plan.

The global energy transition—the core of the wave of climate solutions that can end the crisis—is not about a few grim, determined years of toil and command-and-controlled industrial roar against a backdrop of bloody carnage and constant fear. It's not the desperate spirit of the Dunkirk rescue, not Churchill's call to never yield or Roosevelt's rallying cry after a day that will live in infamy, not even the solidarity of the homefront factories manufacturing vehicles and planes and bombs. No one wants to live like that indefinitely, not if any other choice is on offer, even that of a status quo that will send the whole enterprise lurching closer to the cliff's edge. Reconfiguring the entire world's industrial base will take decades. It will require incentives more enticing and durable than what might be sufficient to muscle through a short march to victory. The transition will be far more analogous to the postwar spirit of renewal—the peace secured, the soldiers returned, GI bills and baby booms, expansion and reconstruction after a catastrophe finally ended in some places and narrowly averted in others. More than anything, the engine of the global energy transition is best understood as a Green Marshall Plan.

I'll run through the history first, as concisely as I can.

In the unstable aftermath of the Second World War, the government of American president Harry S. Truman enlisted recently retired general George C. Marshall, who'd overseen the victorious Allied ground forces, to serve as secretary of state and help negotiate a durable new order for Europe. Abandoning the Soviet Union as an ally, Marshall focused instead on brokering a deal between sixteen Western European countries. The agreement, which was mostly about "economic cooperation," was reached in the fall of 1947 and became a catalyst for the process that would lead to the creation of the European Union in 1993. But more than that, Marshall formulated an audacious plan for a massive American economic aid package, intended to kickstart the reconstruction of the European economy. At a time when many Americans felt they'd already given more than enough—including hundreds of thousands of American lives—to the settling of conflicts in Europe, Marshall called on them to send billions of their tax dollars directly to European countries, including Germany and Italy, whose soldiers had recently been shooting and killing Americans.

The Marshall Plan was far from purely altruistic. It was pitched to skeptical American conservatives in particular as an expensive but necessary defence against rapidly encroaching communism and Soviet dominance. Still, it was an extremely tough sell. Addressing Congress as the European Recovery Plan was unveiled in November 1947, President Truman spoke to an aspect of

America's most flattering sense of itself that will echo in contemporary ears as the motto of the Marvel superhero Spider-Man:

> The American people are becoming more and more deeply aware of their world position. They are learning that great responsibility goes with great power. Our people know that our influence in the world gives us an opportunity— unmatched in history—to conduct ourselves in such a manner that men and women of all the world can move out of the shadows of fear and war and into the light of freedom and peace. We must make the most of that opportunity.

What followed had no precedent in the annals of war, and it seems in retrospect even more staggering in its improbability and scope. Over the next four years the United States government paid for and oversaw the shipment of more than $13 billion in goods, services and expertise to the sixteen participating European countries (the equivalent today would be more than $100 billion). American ships brought food, animal feed and fertilizer, manufactured goods and raw materials, industrial equipment and machinery. At any given moment during the years the Marshall Plan fed Europe's reconstruction, 150 vessels were on their way to or from the continent. They took wheat to Dutch bakers, machinery to the fishing docks of Iceland, tractors for farmers in Denmark, building materials and technical expertise for the construction of new roads and power lines in Greece. American boats delivered barrels of industrial chemicals, tons of sulphur, bales of cotton, truck tires, even soup for schoolchildren in destitute Germany, all packed in cases and sacks marked with the Marshall Plan's red-white-and-blue shield, which read "For European Recovery—Supplied by the United States of America." Future

German chancellor Helmut Kohl remembered a truck doling out soup in his schoolyard under that logo as a formative experience. The Plan's praises were sung in radio broadcasts and pamphlets from Britain to Italy, and the US government paid for those too.

The Marshall Plan was, in the estimation of both its American funders and its European recipients, a roaring success. In Germany, which had seemed almost beyond repair amid the rubble at war's end, it helped trigger what even as early as 1950 was being called *das Wirtschaftswunder* ("the economic miracle"). Truman once said it was "perhaps the greatest venture in constructive statesmanship that any nation has undertaken." A crucial aspect of the Marshall Plan was that its architects, from Marshall on down, insisted that in the end the recovery had to be led by the European governments themselves and must encourage the redevelopment of local skills in local factories. The intention was not to hand out American largesse to a dependent and supplicant Europe, but for Europe to raise itself to the level America had reached through its ferocious wartime economic expansion. Whatever its motivations, the US government saw the intrinsic value of free, prosperous democracies across the Continent.

This is why the Marshall Plan is the stronger metaphor for the solution-driven response the climate crisis requires. War efforts are inherently nationalistic and often nationalized—controlled by individual governments for the exclusive benefit of one country alone. The global energy transition, by contrast, needs innovations from as many sources as possible to create an arsenal of climate solutions that serve everyone's emissions-cutting efforts. For the work of the transition's vanguard to matter, the rest of the world must follow. To deliver on the promise of a much better world, the technologies and techniques of emissions-free industry must spread widely enough to shrink the whole world's carbon footprint as near to zero as possible, as fast as possible. No one nation could possibly do all the work, and duplicating the process over and over again in one jurisdiction after another would be wildly inefficient. Americans drove M4 Sherman tanks, the Soviets fought in T-34s, and Germany had battalions of Panzer IVs. The global energy transition does not need three versions of the same tool.

Consider solar photovoltaic (PV) panels, workhorses of the "new king of electricity" in the 2020s. Solar PV was invented by American researchers at Bell Labs in the 1950s, and solar panels were refined slowly until the 1980s, when Japanese companies like Sharp first developed mass production tools and materials. German manufacturers, generously supported by their government, built on that work as the *Energiewende* raced ahead in the early 2000s. Huge subsidies then shifted much of the mass production to China, which now plays a role in the manufacture of 80 percent of the world's solar panels. The process took decades, and companies around the world are still competing to make panels more efficient, to find materials that might do the job better or more cheaply than the polysilicon in a PV panel, and to look for ways to integrate solar power into a house's roof, a skyscraper's glass window wall, the surface of a road. But in the meantime, the best solar panels currently on the market are readily available for shipping anywhere.

"People will thank Germany in the future for its role." This, as I noted when discussing my 2008 visit to Berlin, was how a German trade official put it to me regarding the country's early, abundant subsidies for the work involved in making solar panels cheap everywhere. It was understood as an effort that would have benefits far beyond Germany's borders. Not at all like a war effort, I mean. This is the way of the global energy transition, and the way it most resembles a Green Marshall Plan. It is crowd-sourced to every level of government and a vast range of industrial sectors, anywhere that sufficient will and investment emerge.

In a moment I'll examine in greater detail the value proposition for much better living to be built by this Green Marshall Plan, but for now understand that its components have largely already been assembled piecemeal, in the same way that Germany and China made solar power cheap and ubiquitous. The government of British Columbia has one of the world's best building codes for accelerating the energy efficiency gains the transition requires. The Canadian government is trying, despite a barrage of political shrapnel, to demonstrate how carbon pricing can be a transformative force in a nation built on resource extraction. Danish urban designers have built one of the world's best toolkits for retrofitting cities for low-carbon living with an enviable quality of life. American urbanists have developed techniques for injecting those innovations into reluctant North American cities. A megalomaniacal entrepreneur in Silicon Valley (born in South Africa and partly educated in Canada, by the way) made electric vehicles the best cars on the road. Then he literally gave away the patents so that everyone might drive an electric car. Automakers from Detroit to Munich to Seoul have joined the effort to make them ubiquitous. The first one most of us will own might well be made in China, though, where electric buses are already commonplace. A physicist who made his emissions-reduction breakthrough just up the hill from my neighbourhood, at the University of Calgary, is now carrying on research at Harvard. He has helped found a company, now based in Squamish, near Vancouver, that might build the first commercially viable machines for pulling carbon dioxide directly out of the atmosphere. These are the technocrats of the Green Marshall Plan.

I think back to the Berlin Energy Transition Dialogue I attended in 2019—as close as the Green Marshall Plan has come to a formal meeting, a gathering of senior government, business and civil society officials at the behest of a government that took the lead in launching the energy transition, for the express purpose of accelerating its uptake by like-minded governments around the world. One of the speakers at the opening session was Joe Kaeser, the CEO of Siemens. "The coming decade," he said, "will definitely be the most dynamic decade in history." He didn't mean just for his industry, because Siemens transcends industries. Siemens is an energy company, I suppose, but not in the sense that it digs coal or drills for oil or even delivers power to homes and business. It's an energy company in the sense that it builds, installs and maintains virtually anything that uses energy. It's been in that business since the company built its first electrical generator to provide power to the world's first system of electric streetlights, in Britain in 1881. And in recent years it has positioned itself as a champion of the global energy transition—serving, for example, as lead sponsor of the City Climate Leadership Awards, an annual competition among the C40 group of major cities around the world pursuing ambitious climate action.

I participated in a Siemens-sponsored "Sustainable Cities Week" junket for international journalists ahead of the 2013 awards ceremony in London. I knew vaguely of Siemens as a big German engineering firm, but I was genuinely awed by the scope of its interests. There were few climate solutions I knew of beyond the reach of the company. They made wind turbines and solar power plant components, locomotives and motors for high-speed trains, subways and light rail systems. In London Siemens was overseeing operation of the city's new congestion-charging system—a vast array of cameras, sensors and digital networks that fed data into an enormous facility on the bank of the Thames. The company was at the same time working on the Crossrail Link, which it touted as "Europe's largest construction project," a simply mammoth reconfiguration of the system of tunnels and stations that move passenger traffic through central London on several national and regional rail systems and multiple tube lines. Siemens was also at work manufacturing more than a thousand new railcars for the higher-capacity train routes it was constructing.

The junket culminated in the presentation of the City Climate Leadership Awards, which took place in a ceremony at a sort of trade show facility called the Crystal that Siemens had erected at the Royal Victoria Dock. It had originally been set up for the 2012 Olympic Games but had become more or less permanent, evolving into a workshop space of sorts—Siemens calls it an "urban development centre"—for the tools the company has developed to help cities install the many technologies of the emissions-free, electric-powered future. The building itself is a green design marvel, of course, but I'd seen plenty of those already. What was striking about it was what it contained.

The Siemens Crystal was essentially a showcase for nearly every tool, test project, green-energy marvel and urban design intervention I'd encountered in a decade on the climate-solutions beat. There it all was, laid out with flair, poise and state-of-the-art user-experience design by one of the world's most influential engineering firms. There were interactive displays, deconstructed models, an elaborate simulator of climate-friendly changes at street level that was like an amped-up version of the *SimCity* video game series. And all of it was branded as the essential toolkit for twenty-first-century urban life. In the lobby outside the Crystal's auditorium, meanwhile, Boris Johnson and Michael Bloomberg made speeches about the necessity and inevitability of an energy transition that had only just begun to adopt that name.

That week in London removed any remaining doubts I had about the inevitability of the transformation I'd been documenting. The global energy transition would eventually spread everywhere, whether delivered by Siemens or by some other enterprising engineering firm that saw a golden business opportunity in the reconfiguration of every city on earth, block by block, to the new specs of an emissions-free world. It was only a question of how fast—and, perhaps, which of those many fine Siemens systems might help do the job in a particular situation. I wonder if that's how it felt to be a young German in the schoolyard, waiting in line as Helmut Kohl once did for a bowl of Marshall Plan soup, finally knowing that a functioning world would soon be rebuilt—and to American standards.

"The belief that the transition is inevitable is the mechanism itself." This was how Laurence Tubiana, a negotiator at the 2015 Paris climate talks, described the outcome five years after the deal was struck. What she meant was that singular sense of momentum that comes from developments like this—the commitment of one of the world's largest engineering firms, currently rebuilding London from its subway tunnels up, to making net zero living work everywhere. And the sense, moreover, that to wait any longer before joining in was to miss out, to be left behind. People respond to that a lot more readily than they do to the panic of a war effort.

Oil, after all, didn't become the world's primary transportation fuel out of panic over the damage done by coal or fear of the dangers encountered on long, dusty trails by rattling horse-drawn wagons. Oil was not a flight from anything. It was a giddy movement toward a sense of real liberation that, at the time, only it could deliver.

3.6 California Dreaming

If you spend any time around urban-planning obsessives—often found gathered under the banner of that full-throated embrace of dense, vibrant, people-centred city living known as urbanism—you're likely to stumble on the writing of James Howard Kunstler. I first encountered Kunstler when I was an undergrad, drawn as a student of twentieth-century American history by the irresistible title of his 1994 magnum opus, *The Geography of Nowhere.* The book is a cultural critique of the suburban model of North American living in the guise of an American jeremiad, a hellfire-and-brimstone condemnation of cul-de-sacs and strip malls and six-lane divided highways.

I'd spent a good part of my childhood lost in a series of suburbias—satellites of St. Louis and Denver (Chesterfield and Aurora, respectively) and a brand-new, curvilinear-streeted, aluminum-sided subdivision in North Bay, a small city built mostly to suburban specs. Kunstler's blistering rant was a welcome balm after all those years spent bored at the mall or freezing in the cold waiting for a city bus. Many years later, at the Congress for the New Urbanism's twentieth annual conference in West Palm Beach, Florida, in 2012, I saw Kunstler deliver his sermon in person. He returned then—as he often does in his writing and speeches—to a line that first appeared in *The Geography of Nowhere.* "Suburbia," Kunstler said, "is best understood as the greatest misallocation of resources in the history of the world."

To Kunstler, the suburban model exists in the realm of self-destructive monuments to human fallibility, somewhere beyond towering sun temples and enormous stone carvings and Roman coliseums feeding bread and circuses to the masses. Suburbia elevates the wasteful spectacle into daily routine—ever larger McMansions on increasingly expansive grass-lawned lots, long boulevards and freeways ushering overgrown SUVs from strip mall to office park to big-box store. Waste as a way of life. And, not incidentally, the most energy-intensive lifestyle the world's ever known, which is to say the most devastating for the planet's climate ever devised. For these and many other reasons, I subscribed for many years after I first read Kunstler to the project of ending suburbia, which was so clearly a catastrophic mass delusion and the engine of our climate-driven demise that there could be no stable future in which it might endure.

It's easy to condemn a Hummer's driver or sneer at a Walmart Supercentre. But Kunstler's glib dismissal of all suburbia as a sort of mass delusion is only useful as a self-congratulatory rhetorical flourish. After all, suburbia is the predominant urban form and preferred residential lifestyle across North America, and in many places beyond. There has to be some explanation for it, some matrix of motivations and incentives that would make sense to Daniel Kahneman and his behavioural economists. And of course there is.

The suburban model was born out of a quest for escape from the crowded, dirty, dangerous cities of the Industrial Revolution's first phase. In the years after the First World War, commuter trains and streetcars and, eventually, the affordably priced private automobile opened up a world of freedom, open space, clean air, tidy streets and lavish gardens. But it was during the prosperous consumerist rush in the years after the Second World War that suburbia's value proposition truly won over the masses. Cheap oil, inexpensive cars, newfangled labour-saving household gadgets— it all added up to an enticing vision of the good life, the fabled American Dream, mass-produced and available to the rapidly expanding middle class, and even to some better-compensated segments of the working class (in cities, for example, where workers made all those cars). The charms of this lifestyle package may have been self-evident to the fast-growing families of the Baby Boom era. But there was one more element that made suburban living not simply a series of household budget decisions but that *little more to life somewhere else* that people had come to yearn for—not just in the United States but across thick swaths of the whole industrialized world. That element was California.

The images, fashions, fads and lifestyle preferences of California in the 1950s and 1960s came to dominate the mass culture of the postwar West so thoroughly that it's easy to forget they were all born of a singular phenomenon. California in those years was the fastest-growing state in the most prosperous nation on earth—Los Angeles alone welcomed nearly 600 new arrivals per day in the late 1950s. California had boomed with hundreds of thousands of jobs making military equipment in the war years, and it kept roaring on afterward. More than that, California was the Western world's dream factory. Hollywood, already the established home of American cinema, became a global production site for movies and television, and the booming California suburbs became its stage set and a trade show pavilion for advertisers selling this new mass culture to the world. Disneyland, McDonald's, the split-level ranch house—these were all born in California in the 1950s. And the meteoric rise of teen culture at the same time was in large measure an idealized image of California teenagers beamed to the world. Malt shops and sock hops, beach parties and surfboards and the Beach Boys crooning at sunset—this was suburbia's most enticing promise of a much better world.

"The state became the postwar age personified—suburbanized, dependent on cars, and the home of the baby boomers." This is

Kirse Granat May, writing in *Golden State, Golden Youth*, an academic study of how California came to dominate the popular culture of those years.

> By 1962, the state claimed the largest population in the nation. Not since the discovery of gold had the state received so much press attention, and the majority of media coverage served to boost the dream. California was America's tomorrowland, in its population growth, suburbs, freeways, lifestyles, and focus on youth. Adding to its power as a magnet in the popular imagination was California's key role in the entertainment industry. This cultural reorientation went beyond the image-making of show business. The modeling of the California family and California youth, a life of cars, fashionable clothing, the drive-in, and the beach, loomed large in the national consciousness.

Along with its expert sales job on the lifestyle, California sold energy-intensive, car-dependent suburban design. How did seemingly half the world buy into that spectacular misallocation of resources that James Howard Kunstler rails against? Listen to a Beach Boys song. Watch a movie starring James Dean or Frankie and Annette. Look at the prizes on TV game shows. (I knew that cars could feature something called "California emissions" long before I had the vaguest clue what the phrase meant, thanks to Bob Barker on *The Price Is Right* in the 1980s.) California packaged and mass-marketed the most enticing value proposition for daily life on offer in the second half of the twentieth century, and it sold with the same speed and ease as McDonald's hamburgers. In my high school years I lived in a bland knock-off of a California split-level ranch house, driving my parents' second car to the mall and the parking lots of fast-food restaurants, playing football and dancing to rock 'n' roll—trying, without ever fully realizing it, to live out some weak caricature of the California dream. And I did so at nearly the other end of the continent, in a harsh northern climate, decades after the Beach Boys first sang odes to T-Birds and little deuce coupes.

A little more to life somewhere else. There is no selling the world on an emissions-free, clean-powered global energy transition without that seductive pitch of a better way. The long prelude of less bad living sold mostly to niche markets. In retail circles in the early days of green products a statistical shorthand emerged to describe the limits of the virtuously less bad option: the thirty-to-three ratio, referring to the 30 percent of consumers who claimed to prefer environmentally friendly products versus the 3 percent who actually spent their money on the green options. A certain number of homeowners (and business owners) would make the effort to put solar panels on the roof or ensure green building standards in their new home or office. A small segment of the car-driving public liked the peace of mind and progressive status of a Toyota Prius or a European-import cargo bike. But the masses have made the sport-utility vehicle parked in a two-car garage on a curving outer-belt suburban street the preferred option for the past twenty years. They're waiting, let's say, on a much better California dream.

The good news is twofold. First, there is the genuinely astounding achievement of that California-born car company Tesla. As the legend has it, the company's founder, Elon Musk, walked away from his executive role at PayPal after it sold to eBay in 2002 with more than $100 million and a drive to solve what he saw as the world's three thorniest problems: viable electric cars, affordable solar power and better space travel. Never mind the rockets; not even two decades later, Tesla has come impressively close to solving the other two. In addition to its cars, which already command the largest share of the electric vehicle market worldwide, Tesla has developed a product line called the Solarglass Roof—a solar panel system that replaces conventional roofing tiles or shingles—and the Powerwall, a household-scale battery designed to store the power from the Solarglass Roof and feed it as needed into the battery packs of a Tesla sedan.

When Tesla unveiled the latest version of the Solarglass Wall in 2019, it sent out a press release that included an artist's mock-up of the full suite of Tesla products. The image depicts a generic modern split-level house, complete with standard two-car garage. The garage's roof is neatly shingled with Tesla's solar tiles. One of its doors stands open, revealing a parked Tesla sedan plugged into a Powerwall mounted on the wall. The message is abundantly clear: *That California-style suburban dream you've all been chasing? We've fully rebooted it for emissions-free twenty-first-century living.* "It'll grow like kelp on steroids," Musk said, with his usual modesty, at the Solarglass Roof product launch.

Now, Elon Musk is—I'll be polite about it—a complicated figure. His belief that luxury tunnels for private cars can solve the traffic problems of the world's cities, or even the idea that cities and transport generally should be centred around the needs of cars—these are deeply misguided and run counter to his avowed commitment to solving the climate crisis. But the good news to take from this snapshot of Tesla's ideal near future is that the tools are pretty much ready to retrofit the status quo for a net zero world.

The other good news—the much better news—is that the energy transition's emerging value proposition does not need to try so hard to mimic the suburban status quo. With more than a century to correct the glaring errors and oversights of a world built to fossil fuel's specs, this new vision does not just swap out dirty energy for clean energy. It delivers a much better world. I've already provided a snapshot of this value proposition. Let's look at it now in more vivid detail.

3.7 The Good Life, Rebooted

Let's start with a house. Because it's best to start somewhere specific, let's start with the first house I ever laid eyes on that generates more energy than it uses. It's a townhouse, one in a unit of six, lined up among five rows of similar units in the German city of Freiburg, their roofs crowned with solar panels and their walls in crayon hues of blue and red and yellow and seafoam green. The townhouse community is called Solarsiedlung ("solar settlement"), and its architect is Rolf Disch. I met Disch and his wife in 2006 and had a friendly chat with them one morning over coffee and miniature Ritter Sport bars in one of the units, which they were using as an office. Disch believed that he'd developed a new design paradigm for sustainable housing, and he wanted to see it duplicated with the same speed and enthusiasm that had spread the Levittown concept of the suburban home around the world half a century earlier.

I returned to Solarsiedlung in 2009 and met with a family who owned one of the townhouses. Because I'd encountered skepticism about the net energy production of the buildings in the interim, I asked to peruse their power bills. They had a stack of papers on file, the bottom lines of which testified that, year after year, their townhouse generated more power than it used. Townhouse as power plant—a simple, revolutionary concept.

So this is where it begins, this value proposition for the good life in the twenty-first century. Homes can be power plants now. Perhaps not all homes at the happy end of this energy transition will make more power than they use, but some will. And many others can produce at least a portion of the energy they require. A few years back, the federal government's Canada Mortgage and Housing Corporation (CMHC) sponsored a nationwide pilot project for net zero housing. This led to the construction of new dwellings in nearly every city in Canada that were designed to at least break even for energy use and production during the course of a year. I toured one in Edmonton that had elongated eaves like a Chinese temple to provide space for all the solar panels. A subdivision in the small city of Okotoks, south of Calgary, has eliminated natural gas use by building a solar-powered district heating system for the entire community, similar to ones I've seen in Europe. This is just in my backyard, more or less. All over the world, the Passive House design standard, born in Canada and refined in Germany, has assisted in promulgating net zero buildings that generate their own energy.

So this is the value proposition's first selling point: *Homes can now be power plants.*

In the best case, the home-as-power-plant is situated in a neighbourhood built for walking, cycling and transit. Vauban, the neighbourhood in which Solarsiedlung is located, is a marvel on this front as well. It was built, as most European neighbourhoods are—and as every urban community was prior to the invention of the automobile—to be easily navigated on foot. All the community's daily needs are within a short walk. The small grocery store doubles as a community hub. Tidy, physically separated bike lanes carry two-wheeled traffic the short ride to Freiburg's full range of amenities. There is also a commuter tramline running straight through the middle of Vauban, with lush grass growing between the tracks, which not only adds to the beauty of the place but muffles the sound of the electric train's motors.

This is the value proposition's second selling point: *Better neighbourhoods built to human scale.*

Many of the world's neighbourhoods built to the scale of California's suburban dreamscape are not so well situated. I've seen vernacular versions of suburbia's spread-out single-family units and curving, overwide streets and big-box retail strips in the UK and Spain, on the outskirts of Bangkok and Hyderabad, in the Dominican Republic and Costa Rica. There is much work to be done to retrofit these communities for life at the scale of a human being on foot. But in the meantime, and because the gentle utopian landscape of a place like Vauban is unlikely to spread everywhere with sufficient speed, there will still be cars, SUVs, pickup trucks. In such places, the energy transition's value proposition promises electrified versions of most of those vehicles, maybe all of them. (Ford's recent introduction of an all-electric version of its bestselling F-150 pickup truck suggests that very few vehicles won't eventually make the switch to electric motors.)

And if not all private vehicles will be all-electric within a decade, more and more of them will be, and more than that, they will be the ones that offer that *little more to life somewhere else.* Here's why: because these neighbourhoods can be retrofitted for smart grids and charging stations much more quickly and easily than they can be rebuilt as compact, idyllic European enclaves. And with those amenities, the value proposition brings this sales pitch: the car as power broker. What do I mean by that? Imagine that electric car—not a fancy luxury Tesla, let's assume a practical Hyundai crossover sedan—parked in a garage, plugged in and ready to charge in the evening. Demand for power is usually lowest at night, and in places with lots of wind power, which often peaks at night, there are often power surpluses. The smart grid allows for real-time pricing, so the car fuels up overnight on the cheap. The owner drives to work the next morning and parks the car in the office park's smart-gridded parking lot, where plugs reconnect the car to the grid. Decide how much of the nearly full battery will be needed for the rest of the day's tasks. Set a boundary—let's say half the charge—and a price threshold. When power demand spikes later in the day, the battery sells power back to the grid, earning the owner a small premium for providing it with a vital energy-storage service. All the technology needed to build this kind of networked grid of renewable energy, real-time pricing, electric vehicles as distributed battery systems, and power brokerage already exists and has been field-tested in places like the island of Bornholm in Denmark.

This is the value proposition's third selling point: *The car as power broker, its fuel as profit.*

Homes and offices, wherever they are located, provide some of the energy transition's greatest opportunities. They are the collective site of what could well become the greatest revolution in energy efficiency the world's ever seen. The homes connected to Bornholm's smart grid have appliances that can make energy-purchase decisions in real time, based on the current price of power and how it's trending. Passive houses and other green buildings worldwide have invented ways to heat, cool, light and power houses, apartment blocks, businesses, schools, factories and warehouses with radically improved efficiency, in any climate. The waste products of one activity are often readily transformed into feedstock for another. In some green neighbourhoods already built, greywater from bathing and cleaning is often used to irrigate gardens, and a community's landfill can become the fuel tank for a biogas power plant (the City of Edmonton does this with its residential waste).

This is the value proposition's fourth selling point: *Efficiency will make the good life more affordable.*

There's much more to it, of course. An efficient home or apartment building that generates at least some of its own energy, in a community where electrified and other emissions-free transportation is plentiful and amenities near at hand, would certainly go a long way toward shrinking the emissions from any individual's daily routine. But there's still a wide, fossil-fuelled world out there, producing about 50 gigatonnes of greenhouse gases per year, and some of the biggest contributors to that tally are industrial energy use (24 percent), agriculture and forestry (18 percent), and cement and chemical production (5 percent). Obviously not everyone will have the means or the interest to find themselves some dense, walkable urban neighbourhood to live and work in. But the reason I find this value proposition compelling is because it flips the climate-solutions conversation. Instead of talking about a disaster to be averted, a crisis to flee from—a *shifting away* from a dire situation into one that is marginally less bad—this value proposition speaks to an opportunity to be seized, a more desirable way to do things, an enhanced quality of life—a *moving toward* a much better world.

More than any of this, though, the value proposition for the good life in the twenty-first century is real. It's tangible. It's as near at hand as a 1950s California ranch house with a nice yard, a two-car garage and a wide freeway nearby. You could pay a visit there next week, as easily as taking a trip to Disneyland.

3.8 Life in Energy-Transition Disneyland

Let's return once more to the Danish island of Bornholm. I was already familiar with the place when I landed there in April 2019—this was my second visit. I knew to catch a public bus at the airport, because even this small agrarian island in the middle of the Baltic Sea has decent public transit. I knew to take a stroll the next morning around the largest town, Rønne, and watch for electric vehicle charging stations, which are on every other block. In the parking lot of a grocery store near the harbour, I found a few Renault Zoe EVs plugged in, part of a local car-sharing service. I knew to rent a bike, because there are safe, physically separated cycle tracks all over the island. And I knew that even though most of Bornholm resembles any other cluster of small Danish towns—old cobblestone squares, half-timbered houses, ochre-shingled roofs, practical modern design—there are many less conspicuous details that set it apart.

Here's what else I knew about Bornholm: Like much of Denmark, its electricity grid relied heavily on renewable energy—thirty-five wind turbines and hundreds of small solar installations scattered across the island supplied half of the island's power. Its remaining electricity and heating needs came from burning wood chips and straw (accounting for 80 percent of its heating), biogas, and a cable connecting its grid to a network in Sweden, which is the nearest mainland. Owing to its high level of intermittent wind power, Bornholm had been selected more than a decade earlier as the site for Denmark's (and probably the world's) most ambitious experiment in next-generation grid technology. From 2013 to 2020, a project in two phases—EcoGrid 1.0 and 2.0—used 2,800 households across Bornholm as a proving ground for smart grids, heat pumps, battery systems and distributed power storage in electric vehicles. More than anything else, though, what set the EcoGrid project apart was its pioneering use of demand response.

So what, exactly, is "demand response"? Well, there's a whole, mostly hidden universe of engineering work and technical gear that goes into delivering consumers power on demand at a steady flow, or frequency, usually 50 hertz in Europe. The grid is very sensitive to changes in frequency that result from inequalities between supply and demand. On a conventional power grid fed by a handful of large power plants, the flow is usually balanced by tweaking supply up or down in response to consumer demand. Grid operators sometimes make deals with large industrial customers to adjust their demand when needed to keep the grid stable. But on a grid fed by renewable energy, the supply of which depends on constant changes in the amount of available wind and sunlight, there is a much larger role for individual consumer demand to play in balancing the grid. This is demand response.

On Bornholm, what demand response meant was that the 2,000 homes in the first phase of EcoGrid shifted their heating and cooling to a temperature range rather than a single trigger temperature, agreeing to draw power for the tasks when it was most abundant. Those homes also had appliances that used smart-grid connections to check the current and trending prices of electricity in order to both reduce costs and shift around energy demand so that the grid could better balance its loads.

One of the lead researchers on the project, Jacob Østergaard, of the Danish Technical University, once broke it down for me with the example of a standard refrigerator. As the fridge's owner, you don't care how or when it keeps things cold, only that the food and drink inside remain within an acceptable range of coolness. So what if, instead of hitting a threshold temperature and immediately demanding power from the grid, the fridge checked with the grid to see where demand and costs were and where they were heading (in the EcoGrid test, electricity was priced in five-minute increments). If demand was high but expected to decline in a couple of hours, the refrigerator could wait awhile and save you a bit of money in the process.

In EcoGrid's second phase, 800 homeowners used heat pumps and electric heat instead of relying on the large-scale district heating systems most common in Bornholm and across Denmark, which heat whole neighbourhoods, often using ultra-high-efficiency furnaces that burn straw or wood waste. The electric heating systems could be controlled remotely, allowing homeowners to reduce waste (by not heating the house when no one was in it) and save money (by heating when electricity was cheaper). On average, energy use in the cold winter months dropped by 30 percent.

The EcoGrid tests inspired the whole island to rethink its place in the world. Bornholm had until recently been a sleepy sort of place with an aging and declining population, getting by on the summer tourist trade from Copenhagen and Sweden. In response to the EcoGrid plans, Bornholm rebranded itself the "Bright Green Island" and began to court smart-grid experiments and other future-tense climate solutions from around the world. Electric car companies came to run tests on their fleets, and so did makers of electric vehicle charging stations. The lovely little craft brewery in the picturesque town of Svaneke started using electric vehicles at its warehouse. The major German energy company E.ON was building its largest Baltic Sea wind farm just off-shore. An elegant inn on the road south along the seashore out of Rønne went by the name of Green Solution House and used solar power to light its rooms, algae to clean its wastewater, and kitchen waste to grow its own vegetables. It was like that at every other turn on Bornholm. I came to think of it as Energy-Transition Disneyland, an immersive model of the better world to come.

One day I biked out to a sort of southern suburb of Rønne called Løkken, a cluster of tidy modern bungalows in dense rows on dead-end streets. Amid the rooftop PV arrays I saw glassed-in gazebos and backyard firepits, lawnmowers and trampolines, breezeways with rows of bikes alongside the parked cars. It was Danish suburbia—Energy-Transition Disneyland's quietly reassuring version of the status quo. Dozens of those houses had participated in possibly the most advanced smart-grid experiment on earth—full exposure to the *terra incognita* of the twenty-first century's radical energy transformation—and they were clearly carrying on just fine.

On another bike ride I grunted up a hill to the theme park's Renewable Energy World—first stopping in at Rønne's Griffen Hotel, where plugged-in electric vehicles in the parking lot were being used to balance intermittent winds and maintain steady power flows. Then I pedalled on to a small community on the edge of town, practically in the shadow of one of the small wind farms supplying the power that needed balancing. I ate a takeout lunch one day on a spit near the main passenger dock, where big ferries from Sweden and Germany arrive daily. I laid out my sausage and roll on a flat rock amid beached fishing boats, gazing out over the dark Baltic, my face whipped by the strong, steady breeze that made the place so attractive a site for wind power. I didn't fully realize it at the time, but I was visiting Bornholm's Land of Giants. Not long after my stay, the Danish government and Ørsted, the country's largest energy company, unveiled plans to build one of the world's largest offshore wind farms, on a shallow shoal called Rønne Banke, just beyond the horizon.

What I mean to say is that Bornholm is cozy and quiet, unassuming, inviting in that often slightly gruff, we're-all-in-this-together way of small-town Denmark. (As I was waiting for the bus back to the airport at the end of my visit, an older woman walking past ordered me in sharp Danish to move my damn suitcase off the narrow sidewalk, a typical no-bullshit approach to community spirit.) It's an easy analogue to the Main Street, USA, that greets Disneyland visitors, a reminder that, for all the fantasy and flash, everyone loves a stroll downtown—or a bicycle ride along the sea-shore. Except on Bornholm it's not nostalgia for a past largely erased by oil-powered automotive progress, but rather a glimpse of a future in which progress is perhaps less intrusive on human-scale daily life.

The gentle ease of life on Bornholm reminded me that discussions of the global energy transition inevitably seem to turn back to the costs and risks. The costs are usually presumed to be steep, and the risks created by the sheer scale of disruption and the myriad proposed changes to the status quo presumed to be perilously high. This is classic Kahneman territory—the aversion to loss is ferociously strong. But what would the full costs of the twentieth-century's value proposition have looked like, had the planners of the California suburban dream laid them out in full on the sidewalk of Main Street, USA? Would cheap gas and split-levels on large lots have looked so enticing if they'd been itemized alongside the gutting of urban neighbourhoods and decimation of small towns, the habituation to rush-hour traffic jams, the constant threat of lethal car crashes, the choking exhaust-pipe smog, the toxic air and soil around the refineries, the incalculable damage of oil spills? And all that even before we get into a full-cost accounting of the climate crisis. The price tag on staying the course is not zero—and it never was.

Contrast all that cost to the vibe on the streets in a Danish town on a small island where the energy transition has most fully laid out its value proposition. Homes on Bornholm (some of them) are power plants. The neighbourhoods were already built to human scale and remain that way. The technology is in place to allow electric cars to serve as power brokers, selling their stored energy back to the grid as a service. Efficiency—which the Danes have excelled at since the energy crisis of the early 1970s taught them with ruthless clarity the full price of imported oil—has made this good life affordable to the average Dane. And this is all real, more real than Disneyland's Main Street, USA, ever was.

There is even a sense of purpose embedded in the fact that Energy-Transition Disneyland is not an escape from daily life but a solution to the existential crisis facing daily life in the twenty-first century. I caught a glimpse of that when I rode my trusty rental bike down to the harbour one morning, to the headquarters of Bornholms Energi og Forsyning (BEOF), where the island's twenty-first-century grid is managed. It was the last day of work before a holiday weekend, and the only engineer there to answer my queries was one of the youngest, a computer scientist named Jeppe Eimose Waagstein. He'd been raised on Bornholm but had gone off to Copenhagen to learn his trade. He'd returned recently, enticed by the prospect of working on the development of such vital components of the energy transition.

Waagstein walked me through the capabilities of some of the latest smart-grid technology on display at the BEOF office, much of it manufactured by Siemens. In addition to the demand-response stuff, there was a test project running as part of EcoGrid 2.0 called ACES (Across Continents Electric Vehicle Services) that I was curious about. Waagstein explained that his office had worked on the project with Nissan, the Danish Technical University, and a manufacturer of electric vehicle charging stations called Nuvve. In addition to the twenty Nissan EVs involved in a physical test on the island, they'd also modelled the operation of Bornholm's grid with full-scale EV penetration into the local market. If, in ten years, the majority of cars on the island were electric, how would that work? It turned out that the frequency-control service those EVs could provide was very valuable indeed—the project's final report estimated that their owners could earn more than €1,000 per year while their cars were parked and plugged into the grid.

In my experience it's exceedingly difficult to get Danes to brag. I spent a few years, for example, trying to find someone who would claim to have formulated any significant part of Copenhagen's world-leading cycling strategy, with one official after another insisting it had just sort of happened by committee. Anyway, that made what Waagstein told me as we were wrapping up our discussion all the more remarkable. He'd come home to Bornholm, he said, because he wanted to help people imagine a better world. "There's a lot of fear of the future. That's why we need to do this." There was such a calm frankness and pride to it, a clarity. I tried repeatedly to follow up with Waagstein in the months after I left Bornholm, hoping he'd tell me more about what it meant to him to be working on the world's most ambitious smart-grid projects. He never replied, which seems to me a very Danish response. The work speaks for itself.

I can imagine what you might be thinking. To paraphrase that great American philosopher Ferris Bueller, *You're not European and you're not planning on being European, so who gives a crap if they're smart-grid innovators?* This is another appealing aspect of Energy-Transition Disneyland: its rides and attractions are portable. They can be transported, installed elsewhere. Consider the Hawaiian island of Kauai—home of that ostentatious luxury home with solar panels and Tesla battery wall that I gawked at from afar during the darkest months of the pandemic—where certain constituent pieces of Bornholm's next-generation grid are being used to transform the island's utility into a North American pacesetter for the transition.

Kauai had long been one of the most expensive electricity markets in the United States, relying on imported oil to fire the turbines of its power stations. In 2008 the island's electricity utility, Kauai Island Utility Cooperative (KIUC), set an ambitious goal of switching to 50 percent renewable power by 2023. Then prices spent the next decade in freefall, and by 2019 the share of renewable power (mostly solar) on the island's grid reached 56 percent. As in Bornholm, KIUC found itself needing energy storage for all that intermittent solar electricity. Bornholm, however, had prioritized using the batteries in EVs, because when its EcoGrid plans first launched, stand-alone energy storage looked far too expensive to do the job by the 2020s. But then storage technologies spent the next decade plummeting in price as well, so Kauai is now home to one of the world's largest solar-plus-storage facilities.

On the site of a former sugar plantation, the Lawai Solar and Energy Storage Project is a 28-megawatt solar farm matched with a commercial-scale 100-megawatt battery storage system, which can hold on to the electricity generated by the day's Hawaiian heat to supply up to six hours of power for evening and early-morning demand. A flock of 300 sheep keep the island's fast-growing vegetation from casting shadows over the panels. For thirty-two straight hours in late 2019, Kauai's electricity was supplied entirely by renewable sources. The Hawaiian Electric Company, which powers the more populous islands of Maui, Oahu and Hawaii, has recently launched plans for solar and storage facilities that will expand statewide capacity tenfold over Kauai's pioneering effort. The Hawaiian approach lacks the demand response and EV integration of Bornholm, but it looks to be racing ahead much more quickly.

As new as the technology, and especially its integration, is—the Bornholm EcoGrid project, after all, was the first of its kind in the world, and it wrapped up only in the fall of 2020—other examples are already emerging. Unsurprisingly, Denmark itself has invested in expanding the reach of smart-grid technology, with a research facility in Copenhagen to bring it to the big city. And EcoGrid's main corporate partner, Siemens, has been conducting smart-grid research in New Brunswick for years. It recently announced a partnership with NB Power and Nova Scotia Power to launch a pilot project called Smart Grid Atlantic, which will test digital tools for use in decentralized grids where consumers are more likely to be directly involved in managing their own demand, and possibly small-scale power producers as well.

When the Danes first embarked on the Bornholm experiment in the mid-aughts, the goal was to design a grid for around 2030. By that point Denmark now expects to draw more than half its energy—not just electricity, but all the energy Danes use every day—from renewable sources, while anticipating that there will be more than a million EVs on its roads. Ten years ago, both of those goals read like open questions. Were they even possible in any technical or practical sense? Now both seem well within reach. As of 2019, 47 percent of the electricity on the Danish grid came from wind power, and about two-thirds of all Danish homes were being heated by district energy systems fed by renewable sources. A handful of Danish islands—most notably Samsø and Aerø, where I first encountered real ambition on the climate-solutions beat back in 2005—have already achieved net zero emissions goals. And then there's Bornholm, where the value proposition for the good life in the twenty-first century was first laid out in full.

The lesson is not that everyone in the world will live like a Bornholmer in 2030, or even in 2050, any more than the whole world now lives in a Los Angeles suburb and drives to the beach every day in a VW Bug with surfboards on the roof. The lesson is that a tangible example exists, a real-life vision, the actual working Main Street, USA, of an Energy-Transition Disneyland. There's a sales brochure now, and it's enticing as hell. To be honest, it's possibly not even the best sales pitch on offer, in Denmark or anywhere else.

As an Aside: New Wonders of the World

It's worth a quick pause here to marvel at the staggering scale of climate solutions in the much better phase now under way. In 2020, for example, Siemens Gamesa unveiled the world's largest wind turbine. The model name is 14-222, which tells the story in shorthand—the new turbine has a 14-megawatt capacity, roughly fourteen times as powerful as the first turbines I ever laid eyes on, back in 2005, and the diameter of its spinning blades is 222 metres. The 14-222 is designed for offshore use, and it will measure about 250 metres from base to hub when standing in the sea. As each blade reaches the apex of its spin, the distance from its tip to the sea floor will be about 360 metres; the very top of the Eiffel Tower's exclamatory finial is 324 metres above the green lawns of the Champs de Mars below.

So imagine an Eiffel Tower standing in choppy waters, and imagine a great rotor mounted to its front like a clock face and spinning in the wind. Now imagine a whole phalanx of them, dozens of them in neat rows. And finally imagine a vast warehouse space on some nearby dock where the machines are manufactured one after the other, like Model Ts rolling down the factory line, Wonders of the World in mass production.

I once visited a serial production site for offshore wind turbines in Bremerhaven, Germany, and that was the idea that immediately leapt to mind—that I was watching the equivalent of an assembly line for Eiffel Towers. Obviously offshore wind's awesome scale isn't all that pleasing by the standards of small-is-beautiful environmentalism, and the humble solar panel remains at least as valuable as a tool in the climate fight. But I find it encouraging nonetheless to see the cleantech industries achieve such awesome scale—a physical manifestation of the level of ambition the task requires.

3.9 The Better Urban Life

When I want to experience daily life in a functional utopia, where the emissions I generate are less than one-fifth the average at home in Calgary, here's what I do: I look for any excuse to visit Copenhagen. After that energy transition conference in Berlin in April 2019, for example, I found a cheap flight to Kastrup airport. (I prefer to take the train, but the schedules didn't quite work for the time I had available to squeeze the trip into.) When I landed, I caught the next metro train headed into the city and made my way to the stop near my friend Mikael's flat. I dropped my bags at his place, borrowed one of his bikes, and headed out on the town, knowing I would have no need of a gasoline-burning engine for any reason whatsoever for the rest of my stay.

Usually on the first available evening after I arrive, Mikael and I ride ten minutes to his favourite wine bar. We bike back much later. Copenhagen's safe, separated bike lanes run virtually everywhere, alongside every major street and many secondary ones. At all hours of the day and night there's at least as much traffic in the bike lanes as in the motor vehicle lanes. Because the lanes are physically separated by a curb and, often, a row of parked cars, and because the signage and lighting has been designed with the usual Danish eye for efficiency, practicality and safety, the ride is as safe as walking, even after a few glasses of the house red.

What those glasses of red always do to me, though, is snap my brain back into Canadian urban cycling mode—tense and hyper-aware, watching for threats from all directions. And that triggers what I've come to think of as my Copenhagen arrival-night ritual: Mikael and I are pedalling slowly back to his flat late at night, the traffic thin, and he does the usual Copenhagen thing, which is to slow down and pull over to one side of the bike lane so I can catch up and we can ride side by side in easy conversation. But I inevitably slow down along with him, thinking of the risks everywhere, thinking I must stay in single file and keep my guard up. And then he realizes what's happening and turns with exaggerated Danish exasperation to holler at me to quit being such a Canadian and get on up here. (Mikael was raised by Danish immigrants in Calgary before he repatriated himself, so he knows both sides of the situation intimately.)

I first met Mikael in 2009. I arrived in Copenhagen keen to learn how the city had come to encompass this impossible retro-futuristic bicycle world, so I rented one of the lovely locally made cargo bikes, piled my kids into the big front bin in the local style, and went to meet Mikael. His full name is Mikael Colville-Andersen, by the way, and he will need little further introduction for urban cycling enthusiasts. His fondness for photographing stylish Copenhageners on their bikes and posting the pictures on a blog he gave the grand title "Copenhagen Cycle Chic" turned him, first, into one of the world's foremost spokespeople for urban cycling the Danish way, and then into a design consultant and speaker in high international demand.

As interest in urban cycling grew, the urban-design world came quickly to understand that there was no big city with better cycling infrastructure than Copenhagen's. (The only real rival is Amsterdam. I'd call it a draw between equals competing in an entire separate Champions' League from the rest of the world.) And there was no better way to learn how it worked than to take up Mikael on an offer to lend you a bike and lead you on a tour of the city. Over the years he has conducted tours for foreign politicians, planners and journalists from seemingly every major city on earth.

There are many delightful details along the way in Mikael's speech from the bicycle seat—I've heard several versions of it in the years since the first time, often laced with his trademark profanity—but the primary answer to how Copenhagen became a global cycling capital is simplicity itself. Bikes work in Copenhagen because they are the simplest, fastest, most convenient and most pleasant way to get around. They are automatic, the default setting— the way backing an SUV out of the garage is for residents of North American suburbia. About half of Copenhagen's residents (and an even larger share of downtown residents) commute primarily by bike, and they routinely report that their two main reasons for doing so are because it's the quickest and most convenient way to get around. Smaller numbers say they like the exercise or want to save money. Only 7 percent say they commute by bike because it's "eco-friendly." This is one of the great surprises of the Copenhagen model of urban design—it was not devised primarily to reduce emissions and save the planet. It was not the result of an activist crusade or political campaign. Copenhageners stumbled by accident on a core principle of urban sustainability: the tools that enhance quality of life in cities are the same ones that slash emissions. Copenhagen became a world leader in urban climate solutions as a sort of incidental factor on the way to becoming a much better city for people to live in.

There was no grand plan, no thundering declaration of intent. No one announced that in a generation Copenhagen would become the world's foremost cycling city and a perennial contender for the title in global rankings of the world's most livable and sustainable cities. It simply reverted to a social order in which people and their daily needs took priority over the flow and parking of motor vehicle traffic, and much of the rest followed. Because the city remains a dense European capital of mid-rise apartment blocks and mixed-use buildings, it was readily converted to more efficient district heating. When Denmark began its push into renewables, an offshore wind farm went up in the harbour and the coal and gas plants in the region converted to renewable biomass (wood chips and straw, mostly). By the time the rest of the world had awakened to the scope of the climate crisis, Copenhagen was already far down its path to the goal.

Perhaps the very best news out of Copenhagen is that its model for better urban life is readily exportable. In fact, it's already been exported, in bits and pieces. Many of the first such exports were made under the auspices of a Danish architect and urban designer named Jan Gehl, whose transformative insight into urban design the Copenhagen way began with counting people and what they were doing in the inner city after the streets started closing to cars. He counted how many sat at cafés, how many were strolling down the Strøget, how many lingered next to this fountain or on that bench. Then he took his data and built a sort of toolkit for retrofitting city streets for people, and he found clients keen to use it the world over. He sometimes took to calling it "reconquest"—a liberation of the twenty-first-century city from the tyranny of car dependency.

Gehl's consultations led the city of Melbourne, Australia, to discover that the dowdy old laneways between its inner-city buildings could quickly become a snaking network of small shops and outdoor cafés. What was once called "an empty, soulless city centre" by a local newspaper became the centrepiece of one of the world's most livable cities. In New York under Mayor Michael Bloomberg and his visionary transportation planner, Janette Sadik-Khan, Gehl's Copenhagen model inspired the closing of Manhattan plazas—including iconic Times Square—to vehicle traffic, an overhaul of the pedestrian realm on Broadway, the addition of more than 300 kilometres of bike lanes, and a rethinking of street life throughout all five boroughs of the city.

Gehl's approach—which is to say Copenhagen's—has been copied with similar success in London and Oslo, Denver and Vancouver, São Paolo, Brazil, and Guangzhou, China. Bike lanes and better sidewalks do not by themselves erase emissions from coal plants and cars, of course, but they outfit cities with the street-level infrastructure needed to both shrink the energy needs of the average resident and more readily shift to emissions-free living. A suburban cul-de-sac with solar panels on every roof is still hopelessly energy-intensive; an urban neighbourhood with complete streets is the basic building block of sustainable city living.

Again, though, I hear echoes of Ferris Bueller. *We're not European, and we don't plan on being European.* We don't all live in New York or London either. Fair enough. But I live in Calgary—a city not only deeply car-centred in its design but built on the proceeds of oil money to cater to the needs of workers in the oil industry—and I've seen the inner city's first physically separated bike lanes on downtown streets increase commuting by bicycle by more than 50 percent in their first three years alone. (To be fair, far more Calgarians commute on North America's most heavily used LRT system.) Visiting family in Nova Scotia, I've watched as the seasonal sidewalk widening of one block in downtown Halifax has grown into a network of pedestrian and pedestrian-priority streets, readily transforming a handful of side streets into the very heart of the city.

And how about this? I once spent three months living in a small town within an afternoon's drive of the Arctic Circle, without a car and without hassle, and the same urban design principles applied. The town was Dawson City, Yukon. Because it's half-frozen in time during the Klondike Gold Rush, ten years before Ford built his first

Model T, Dawson City is built to the same human scale as down-town Copenhagen. I never once felt troubled by the absence of a car. Did I mention I was there from January to March? Some mornings you could watch the mercury in the thermometer outside the kitchen window vanish into the little bulb at the bottom, as the temperature dropped well below minus 40°C and the forecast called for "ice fog." Even in Dawson, where snaking extension cords tether a vehicle's block heater to an electrical socket, cars may not start at that temperature. I put on my warmest clothes and ran an errand on foot that day, just to prove that I could.

If Copenhagen-style urbanism works in the Yukon in February, surely it could work anywhere. It's really just a matter of giving people a taste of it.

3.10 Much Better Blocks

I first heard the term "tactical urbanism" (and of its transforma-tive potential) at the twentieth Congress for the New Urbanism (CNU20) in West Palm Beach, Florida, in 2012. The big keynotes and plenary sessions at the conference focused mostly on big-budget, large-scale projects by the new urbanism's established stars, figures such as Andrés Duany and Peter Calthorpe and Ellen Dunham-Jones. That first wave of new urbanists had spent the previous twenty years carving out a significant niche, building new communities and urban developments that looked more like old Copenhagen than suburban Los Angeles.

These days, nearly every big city (and many smaller ones) in North America has a new-urbanist community or two. There's one here in Calgary called McKenzie Towne, with a makeshift main street and homes with front porches. In suburban Toronto there's a community called Cornell. A dying shopping mall in the Denver suburb of Lakewood was redeveloped as an old-fashioned downtown and rechristened Belmar. These are just the ones I've visited myself, but there are dozens of others, and all of them take pains to bring human-scale walkability and a greater mix of urban amenities to new suburbs.

The new urbanism was a vital reboot of the city-building conversation in North America, and its work has laid some of the foundations necessary for urban climate solutions. But by CNU20 I felt as if I'd heard many of those stories before. What truly excited me was a presentation at one of the smaller sessions, by a young new urbanist named Mike Lydon who was working on a concept he had dubbed "tactical urbanism." What he meant by that was nimble, cheap, small-scale changes to the fabric of a city street, often temporary, that could demonstrate how readily an urban space can be repurposed and improved to catalyze greater change. The PARK(ing) Day movement, which launched in 2005 in San Francisco with the one-day reclamation of a single street-side parking space as a tiny temporary city park and then spread to dozens of other cities— that's tactical urbanism. So are temporary sidewalk expansions to create wider walkways and café spaces, guerilla installations of traffic-calming measures, and the proliferation of food trucks.

"Because these projects tend to be on a much smaller scale, they're easier to replicate," Lydon told me.

They're manageable. It's hard to plan and build a bridge or a new light rail line without a ten-year process, funding, approvals, official permits, and so on. But if you're going to make a small-scale change to an intersection and it works really well, then guess what? You might have a hundred more of those intersections around town. And that's how you get a much wider, more systemic change in the city.

Maybe the strongest example of transformative tactical urbanism I've encountered over the past ten years has been the "Better Block" movement. It began, like most tactical urbanist campaigns, with a single simple idea in one specific place. In Dallas, Texas, in the neighbourhood of Oak Cliff, a handful of local entrepreneurs, artists and activists wanted to make their main street better. Oak Cliff had once been a vibrant streetcar suburb, but the city had dismantled its streetcar network and pulled up the tracks in 1956, and then it had turned Oak Cliff's shopping and commercial blocks into one-way streets for faster-moving commuter traffic. What's more—and this is always the way with places in need of tactical urbanism—the city's bylaws thwarted every effort to improve the neighbourhood. Probably the biggest barrier in Oak Cliff was the off-street parking requirement. To open a new small business—a café or bookshop, say—you had to provide a minimum number of parking spaces off the street. That meant a bigger lot, which meant much higher start-up costs, which meant that none of the small businesses that can bring life to a city street could afford to launch in Oak Cliff.

The folks there realized they'd never be able to convince City Hall to agree to reverse parking requirements or approve bike lanes or loosen the regulations on outdoor café seating all at once. Instead, the Oak Cliff interventionists—led by Jason Roberts, a local musician, and Andrew Howard, a transportation planner—went to City Hall in the spring of 2010 to obtain a simple special event permit. As part of the Oak Cliff Art Crawl, they would construct a "living block" as an "art installation." They called the project "Build a Better Block."

"It was this punk-rock, stick-it-to-the-man thing," Roberts told the *Houston Chronicle* afterward. "We flouted every law we could." With a budget well under $1,000, borrowed materials and volunteer labour, they turned a single block into a snapshot of a thriving downtown for a single weekend. Thrift-store chairs were set up around hay-bale tables to create a pop-up street café. Planters were installed in rows to create safe bike lanes. There was a children's art studio and a flower shop. It might have been punk rock in tactics, but it was far from radical. It was nice. It was a re-creation of the quiet little streets that North Americans travel to Europe in droves to immerse themselves in. And, more than that, it was *tangible*. This wasn't an argument or a theory or a conceptual plan. It was a place that people could come to and experience—and understand in an instant that their neighbourhood streets could be more than funnels for commuter motor vehicle traffic.

"Instead of town hall meetings, charrettes, and long discussions, just go onsite to where the problem is and start fixing things within days, not years"—this is how Team Better Block's Andrew Howard put it.

Like any self-respecting indie-culture maven circa 2010, Team Better Block was committed to "open source" methods. They wanted anyone who was so motivated to be able to copy what they'd done. So they documented everything, put together an elegant video to show how they'd done it, and threw it up on YouTube. Big surprise—the thing went viral. Better Block installations popped up in nearby Fort Worth and then San Antonio, and then in Saint Paul, Minnesota, Lafayette, Louisiana, and Wichita, Kansas—and then all over the damn place. Memphis. Indianapolis. Portland, Oregon. Barberton, Ohio. Vieux-Saint-Jean, Quebec. Saskatoon. Belfast. (Belfast?) Tehran. (Tehran!) Dozens in a few short years.

An urban world starved for more vibrant streets had simply been waiting for permission to intervene.

"No city will build a bridge or a light-rail system with tactical urbanism alone," Mike Lydon explains. "But creative and smart interventions can build the social and political capital needed to push such projects forward from the study and proposal stage. Tactical urbanism looks physical, but often the best results are social, in building more capacity and ties to longer-term change within neighborhoods."

It's telling that all these initiatives—the new urbanism, Better Blocks, Copenhagen's original model of better urban living—are only incidentally or retroactively about finding climate solutions. In each case, people simply wanted enhanced street life, more vibrant neighbourhoods, better cities. I think back on that twentieth Congress for the New Urbanism. In one session I listened as James Howard Kunstler railed with Old Testament fury against the wasted resources of suburbia, prophesying an apocalyptic end for the whole suburban dream. Kunstler was very much in hair-shirt green mode, anticipating a world of harder work as penance when the cheap energy ran out. Meanwhile, in another session there was Mike Lydon, energetic and hopeful, speaking with dark euphoria about the quick, easy, even joyful ways by which citizens could reclaim their right to the public streets around them.

The tactical urbanists are, in this respect, close kin to the Tesla automotive design model. Making a block (or a city) better is not about reducing a certain percentage of emissions by a particular year. It's not about saving the planet at all, even though it's at least as important for ending the climate crisis as any given protest against pipelines. It's about the opportunity to live a better life, to sit out on the street sipping a glass of wine on a nice evening, watching the pageant of urban life unfold around you. This was stolen from many urban neighbourhoods, however unintentionally, by half a century spent reconfiguring cities for the priorities of the automobile.

I'm reminded of the film *Who Framed Roger Rabbit?*, set in the rapidly expanding Los Angeles of the 1940s. The plot centres on a diabolical plan hatched by the movie's villain, gleefully portrayed by Christopher Lloyd, to level vibrant Toon Town, tear up the streetcar tracks, and melt down cartoon characters into an oily "dip" that would serve as the raw material for freeways. The architects of the real-world transformation (mostly) weren't quite so cacklingly evil, but the results have been the same. The streetcar tracks are gone, the old streets are ersatz one-way freeways, and people have been erased from the picture. The reconquest of the world's cities—a vital necessity for ending the climate crisis—began by simply counting people on streets and watching what they were doing.

It's mostly an easy sell from some angles. Who, after all, would stand against vibrant streets in general? But there's a trick to it, because the force that produces vibrant streets is one that in many circles has long been cast in an unattractive light. That force is density, my friends. Glorious urban density.

3.11 Ode to Unsung Climate Heroes, Part 1: Density

I first fully fell in love with city living in the downtown Toronto neighbourhood surrounding my first bachelor apartment. It was above a small grocer, right across the street from Ossington subway station. The neighbourhood was not hip in those days, but I loved its dive bars and cafés full of Europeans cursing at football matches all the more for it. My apartment was 600 square feet and cockroach infested, but my quality of life had never been higher. I was in the middle of teeming urban life, and two minutes from a subway line that could take me anywhere I wanted in one of the world's great cities.

I came home from class one November afternoon, mounting the sidewalk steps with the nonchalance of daily routine. It had just started to snow, the first snowflakes of approaching winter, big and heavy, muting the traffic noise and seeming to close in the city around me in that magical way snow does. I decided to go up the block to a new bar and grill that had just opened. I'd chatted with the owner a bit the first time I stopped in. He was Iranian and giddy with pride about his place. He told me a story of how he'd been chipping away the old signage on his storefront and had found a sign beneath it, and then a sign beneath that, and he kept digging till he found glorious old etched-glass windows and a faded sign with a horse silhouette on it, and that's why he'd named it the Black Horse Tavern.

It was mid-afternoon and he was trying out a new dish in the kitchen, so he offered me a plate—a quarter chicken in a sauce with pomegranate. He was calling it "Persian chicken," he told me regretfully, because no one thought of fine food when they heard "Iranian." I ate my Persian chicken and drank a beer and bullshit-ted with the bar owner, watching the snow drop its silent blanket on Bloor Street, and I knew that I'd never felt more *alive* on a mundane November afternoon across the street from where I lived.

That was how I learned that a well-located apartment, close to shops and cafés and transit, could be far superior to a house with a front lawn and back deck and two-car garage. I didn't know the name of the urban force that delivered that value then, but I do now. It was urban density—a free service provided by a well-designed neighbourhood in a vibrant city. And it's one of the most important and underappreciated tools in the climate-solutions toolkit.

Maybe it's a naming problem—being called dense isn't a compliment, unless you're a mercury molecule—or maybe it's a long-lingering hangover from the days when inner cities were synonymous with the dirt, noise, danger and disease of heavy industry. In any case, generations of North Americans have been raised to see density not as a gift but a curse. They crowd community meetings and city halls to stop zoning changes that would allow more people to live nearby, and they rail against taller buildings. In some places they even fight against expansion of mass transit into their neighbourhoods (there is a wholly unfounded urban myth that criminals ride subways and LRTs to wealthier parts of town to commit petty crimes and then flee to safety with their pilfered wares on those all-too-convenient conveyances). In reality, of course, proximity to mass transit and greater density bring life and new business to nearby streets and boost property values and quality of life alike. A map of the world's most highly prized (and ludicrously expensive) real estate is mostly an overlay of a map of urban density.

What a bonus, then, that such a desirable tool of better city building is also such an essential climate solution. Done right, density makes all the other tools of urban emissions-cutting much easier and more effective. It is a great enabler, for example, of energy efficiency. All those biking Copenhageners owe their tiny carbon footprints not just to their car-free commutes but also to district heating, which pipes heat from hyper-efficient renewably powered boilers to whole neighbourhoods instead of wastefully generating heat on-site at every building or home. Density makes transit, cycling, walking, car-sharing, bike-sharing, scooter-sharing—every form of transportation other than a private car—easier and cheaper to deliver and much more convenient to access. Density does not change the source of a home's energy or oblige an office building to install more efficient windows, but it does reduce the daily energy needs of the people in those homes and offices. And that turns out to be a substantial climate solution all by itself.

Consider the results of a simple sort of modelling experiment conducted by an Oregon think tank called the Sightline Institute. Imagine two city blocks with eighteen single-family houses on each block. Now imagine that three old houses on each block are torn down and the sites redeveloped. On one block, three 3,000-square-foot single-family homes go up; on the other, the three lots become a duplex, a triplex and a fourplex. Total construction costs would be roughly equal. The institute dubbed these two sites the McMansion Block and the Plex Block.

The McMansion Block represents the more common infill strategy in much of urban North America, especially since many cities and nearly all their suburbs have strict zoning restrictions against multi-unit redevelopments on residential streets. In most major cities in North America, at least half the land does not permit increases in density. (The data on this were surprising even to me, as I'd anticipated they'd be quite high. In San Jose, California, 94 percent of residential areas are zoned for single-family detached homes only, while the figures for Los Angeles, Portland, Seattle and Chicago are all higher than 70 percent. In Vancouver, 81 percent of residential lots are single-family only, and in Toronto the figure is 62 percent.) The Sightline experiment compared the energy needs of the McMansion Block and the Plex Block. They found that overall energy use was 20 percent less per household for the entire Plex Block—not because the existing houses grew more efficient but simply because that's how much less energy a household in a 1,000-square-foot fourplex unit requires.

I like to think of this in terms of my proudly progressive downtown neighbourhood. It's full of hundred-year-old single-family
homes and little mid-century bungalows that are constantly being
bought up and replaced with 3,000-square-foot hulks. It's a neighbourhood full of diligent recyclers and bike commuters and transit users and electric-car buyers. There are several rooftop solar
arrays and backyard beehives within a couple of blocks of my
house alone. There's a weekly community farmers' market and
community vegetable gardens. And yet, as a neighbourhood, we'd
likely do more for our collective emissions simply by becoming
vocal advocates for multi-unit infills on our streets. A study of
urban communities in California, for example, found that in the
city of Berkeley—like my neighbourhood, a place full of well-
meaning owners of single-family homes—the single biggest step
the municipality could take to reduce its emissions by 2030, on a
list that included electric vehicles, renewable energy and building
efficiency, was to legislate more residential density via urban infill.

Density, though, is often a hard sell. It sounds like a burden. It
needs champions. And I'd suggest that those champions should
probably try to think more like tactical urbanists than conventional
urban planners. Begin with the promise of a Better Block. Let
people see and feel and immerse themselves in the benefits of dense,
vibrant urban life. Let them sit out at a café or ride with their kids
down a safe bike lane. Then they might be more amenable to discussions about the hidden value of triplexes up the street.

This is how you get density to play to the biases of the human
mind rather than trying to argue past the powerful force of loss
aversion. In scale and scope, daily life in a suitably dense urban
neighbourhood feels immediately welcoming and completely
comfortable to pretty much everyone. It's so deeply ingrained and
universally appreciated that it inspires layouts of theme parks

with their Main Street, USAs, and feeds the entire European tourist industry. Which is to say that North Americans routinely take vacations *to* urban density. Whether the destination is Disneyland or Paris or New York, they seek out walkable, sustainable urban neighbourhoods for their leisure time. They simply need to be reintroduced to them in their own backyards. And doing so would be a vital climate solution, because it would unlock that other great unsung climate hero: efficiency.

3.12   Ode to Unsung Climate Heroes, Part 2: Efficiency

In 2018 I worked on a report for Natural Resources Canada, summarizing the findings of a roundtable of energy experts charged with developing a plausible vision for Canada's long-term future in a low-carbon economy. A detail in the report that particularly surprised me was that Canada could meet more than 30 percent of its Paris emissions target simply by implementing all the measures contained in the provincial and federal efficiency commitments already on the books. About *one-third* of the vital, elusive emissions cuts required, simply by wasting less! And the IEA reports the worldwide figure is more like 40 percent. That's far more substantial than a vast fleet of Tesla sedans or a solar array on the roof of every other house in your neighbourhood.

Efficiency never gets the headlines, though. It sits there on a dusty shelf alongside foresight and prudence and thrift. Everyone wants the benefits of these virtues, but no one is, you know, *into* them. Like good design generally, it works at its very best when you don't notice it at all. Yet energy efficiency is the very foundation of a much better world. It enhances everyday life and improves the ease of daily routines. It makes navigating the energy transition both cheaper and faster.

Let me walk you through a case in point. I mean that literally. Let's walk through Unilever-Haus, European headquarters of the world's largest producer of consumer goods, in Hamburg, Germany, as I did on one grey afternoon. Unilever-Haus is one of the anchor buildings of the HafenCity district on Hamburg's waterfront, a mammoth multi-year redevelopment of a vast swath of the city's docklands. The district is anchored by the Elbphilharmonie, a new concert hall that became the symbol of twenty-first-century Hamburg the day it opened in 2017.

Unilever-Haus itself is a more modest building, six storeys of angular white metal and glass housing 1,100 Unilever employees and a handful of stores and services on the ground floor. It wouldn't look wildly out of place in any office park in North America. What I mean is, it's a stylish piece of architecture that doesn't leap out as an exemplary one. But when it opened in 2009, it might have been the greenest corporate headquarters on earth.

Unilever-Haus requires 70 percent less energy than the already quite efficient German norm. It was the first office building in the world illuminated entirely by LED lighting. Its design is centred around a soaring central atrium that opens the building to the sun's light and heat. An enormous circular diffuser hanging from the atrium's ceiling bounces natural light into the surrounding offices, reducing the artificial lighting needed, and clever natural ventilation has minimized the building's heating and cooling requirements. It's a marvel of energy efficiency, one of the best such examples on earth on the afternoon in 2011 that one of its architects, Martin Haas of Behnisch Architekten, led me through it.

Haas was delivering his standard spiel to a group of visiting journalists as we toured the building, pointing out an energy-saving feature here and a sustainability feature there. His approach was precise and technical, not a marketing pitch. The main atrium was crisscrossed at various levels by skywalks, and we stopped midway across the highest one while Haas explained that Unilever's design brief hadn't explicitly called for the greenest office building in Europe, or even reaching a particular efficiency goal. It was intended, first and foremost, to be a space that facilitated informal interactions and easy collaboration between office workers. The company's Hamburg offices had previously spread over multiple floors of two buildings. Almost any discussion at all, no matter how incidental, had required a formal meeting. The architects working on the design—a one-off collaboration among a handful of firms based in Stuttgart—had decided to envision the place as a living space instead of a workspace. In place of upscale creature comforts, however, they optimized its environmental performance, recognizing that a climate-friendly building would also be more people-friendly.

That's when it hit me. It was late afternoon, and I'd spent a long week gawking at exemplary green design. I'd been gazing around the space, starting to daydream. The atrium's roof allowed a clear view of the drizzly sky above. It was one of those bland mid-week afternoons when office workers the world over are given to surreptitiously playing games on their computers as they watch the clock tick yawningly toward quitting time. And yet the concourse spaces overlooking the atrium all around us were filled with little knots of Unilever employees in twos and threes, leaning against railings or hunched together over tables. Everyone was lingering in the welcoming space and collaborating informally with such offhand ease that the scene looked staged. The atrium was more comfortable and brighter than the city around it, the air quality inside so fine that the move to the new headquarters had triggered a significant drop in sick days.

The feats of engineering and design that made the atrium so inviting, that made Unilever-Haus among the world's most efficient and least polluting buildings, were largely invisible to the untrained eye. The climate heroics of world-class efficiency are not flashy. But if they can be rendered visible—if they become objects of desire— they are world-beating climate solutions. I'm thinking in particular of a disciple of Daniel Kahneman, a business professor at Arizona State University named Robert Cialdini. Cialdini decided several years back to use the lessons of behavioural economics to encourage more people to participate in energy-efficiency campaigns. Those campaigns have been ubiquitous since at least the energy crisis of the 1970s. Growing up on Canadian military bases, I lived in homes whose walls were adorned with small stickers, usually next to light switches, to encourage us to turn off the lights and turn down the thermostat. "ENERGY EFFICIENCY = COMBAT READINESS," they read. Every power company and corporate cost-cutting department in the free world surely had its own campaign.

When Cialdini decided to investigate how people decide to participate in such campaigns, he chose the most ubiquitous (and possibly the least consequential) climate-solutions campaign yet mounted: the efforts of seemingly every hotel on earth to encourage guests to reuse their towels and linens. Communicated through a wide range of doorknob hangers, side-table stand-up cards, brochures and key-card holders, the existing reuse programs offered a staggering range of messages and incentives. Some hotels talked about the energy saved, others about the emissions eliminated. Some made appeals to cost-efficiency, others to salvation of the whole planet. There were even hotels that offered breakfast discounts or other little bonuses for participating.

In field tests, Cialdini tried a wide range of standard approaches. None was significantly more successful than any other at encouraging greater participation. What did work was carefully executed behavioural economics. If you simply told guests that most of the other guests in the hotel were participating in the program, participation increased by an average of 26 percent. If you told them that most of the other guests *in that very room* had participated, participation increased by 33 percent. Hotel guests, it turned out, weren't looking to make a statement, save the world or win a prize. They simply wanted to do what other people like them were doing.

Cialdini took his findings and helped launch a consulting firm called Opower to work with energy companies on reducing household energy use. The new power bills that Opower introduced in San Diego showed residents how their power use compared to that of other users, using a clear, simple bar graph. The best results came when the comparison was not to the whole state of California or the entire city but rather to the immediate neighbourhood. Using the Opower design, the local utility in San Diego soon saw energy use decline in households whose bills showed them to be using more than the neighbourhood average. There was, however, a catch—households whose energy use was *below* the average often increased their power use in the next billing period. The bar graph had given them permission to slack off. So Opower introduced another highly sophisticated communications tool to discourage recidivism among the more frugal customers: a happy face. That mostly took care of it. In the invisible world of heroic efficiency, even the slightest wink toward visibility and recognition can be powerful stuff.

At its very best, efficiency doesn't need that wink. It becomes the default setting, because it does the job better. And simply doing better is the ultimate goal of the most important energy-efficiency initiatives that have emerged in the past decade, and that will come to dominate the building trades in the next one. These aren't exemplary individual buildings or companies. They are standards, codes, certification systems. Possibly the best-known one is LEED—a set of building construction and performance criteria that became the preferred way to indicate commitment to climate-friendly construction, whether for a company building its headquarters or a city burnishing its green bona fides.

LEED standards and their Olympian silver/gold/platinum certification hierarchy spread around the world rapidly after the US Green Building Council introduced its first comprehensive grading system in 2009. I've toured, worked in and slept in LEED-certified buildings from sunny Florida to frosty Fort McMurray. The first LEED platinum building certified outside the United States was a conference centre in southern India. Apple's Silicon Valley headquarters is certified platinum, and so is Manitoba Hydro's office tower in downtown Winnipeg. The University of North Texas plays its varsity football games in a LEED platinum stadium, and the San Francisco 49ers play their games in a LEED gold one. And on and on—thousands of buildings and more than 100,000 professionals who sport the credentials to conduct certifications.

There are numerous other standards. The Passive House Standard emerged from the Passivhaus Institut in Darmstadt, Germany, and expanded to North America when the Passive House Institute US opened in Chicago, often yielding even more impressive results. On Park Avenue in the Bronx, for example, a solar-powered passive-house apartment building with 154 low-income units opened in 2019, demonstrating unequivocally that high-performance, hyper-efficient design isn't solely for luxury clients and corporate giants. Some European countries, meanwhile, have their own green building certifications, and others don't even bother. When I asked Martin Haas whether Unilever-Haus had received any LEED-type plaudits, he scoffed gently and told me that Germany's baseline building codes exceeded LEED standards. Which is, as I said, the ultimate goal—to make the climate-friendliest option the default. The real value of these certifications is that they serve as catalysts, creating an early-adopter market that in time makes the design innovations ubiquitous.

The world will not rebuild itself completely in the next decade, or the next three. So a huge role for these emerging new standards will be to require the outfitting of old buildings with new windows and better insulation, replacing inefficient appliances, and—maybe most importantly—swapping out old boilers, water heaters and air-conditioning units fed by natural gas and other fossil fuels for electric ones. In fact, let's be grandiose about it. Let's build this efficiency ode into a great climactic (and climatic) aria by singing boundless praises for the humble heat pump.

Yes, the heat pump. Air-source or ground-source—that is, fed either by warm air or the warmth of the earth beneath our feet. A simple technology as old as any given refrigerator or air conditioner, both of which employ heat pumps of different types to compress warm air inside the fridge or from inside the building and force it out, leaving cooler air behind. Oh, how we have neglected our heat pumps, right there in plain sight, as we fuss over rooftop solar water heaters and household-scale battery packs and the like. But the heat pump has waited, patient and unpretentious, quietly stupendous, for its moment to finally arrive. And oh how it has.

To date, you tend to see heat pumps almost anywhere you find hyper-efficient homes. Green offices chasing LEED accolades, net zero houses, low-emissions wonders of all stripes—often their interior heating and cooling are aided by a heat pump. It can either use the differential in temperature between the inside and outside of a structure to heat or cool the interior, or it can tap into the reliably roughly room-temperature earth somewhere between 30 and 100 feet below ground to keep the space above warm in winter and cool in summer. Either way, the heat pump's never a marquee feature of the green building scene. It's the trusty workhorse. And oh, my friends, what work it does!

Worldwide, buildings generate roughly 40 percent of all carbon dioxide emissions, if emissions from construction and materials are also factored in. And one-third or more of those emissions come from heating, cooling and ventilating the space inside. In the United States, indoor heating alone makes up one-tenth of the country's second-largest-on-earth carbon footprint. Swapping out all that fossil-fuelled temperature control for heat pumps can cut emissions by 40 percent or more for the average building. And if those buildings are connected to grids that are predominantly or entirely emissions-free, then the greenhouse gas reduction is anywhere from 70 to 100 percent. Heat pumps are far more efficient than gas boilers—up to four times more efficient—and they generate more heat energy than they consume in electrical energy. And though the upfront cost of a heat pump is at present often a good deal higher than the cost of a conventional gas boiler, the heat pump is far cheaper over its lifetime.

Cheaper, more efficient, a crucial but uncelebrated climate solution—the heat pump is the whole efficiency ode writ small. Climate advocates obsess over the number of long-haul flights they take and the vehicle miles travelled by their greenhouse-ripened tomatoes. But where are the strident marchers calling for a heat pump in every home, the cornerstone of an efficiency revolution that could hack back a third of humanity's emissions all by itself?

With or without fanfare, the efficiency march remains a vital necessity—particularly in order to make it possible, in the years to come, to electrify almost everything.

3.13 The Electric Age

I've mentioned British Columbia's pacesetting Energy Step Code. Let me explain now how I came to be so familiar with one province's somewhat obscure efficiency-minded building-code changes. I have a colleague in Vancouver named James Glave, who works with the same kinds of organizations I do to communicate more clearly the benefits of climate solutions, such as BC's Step Code. James got a tattoo on his arm a few years back, a sort of statement of purpose: "ELECTRIFY EVERYTHING." To shrink emissions as near to zero as possible by mid-century, the world will have to rely much more than ever before on electricity. The entire climate-solving project, from a certain angle, is simply a global effort to electrify everything.

Let's use James's native British Columbia as a case in point. It begins as a hydroelectric powerhouse with close to zero emissions on its grid—more than 95 percent of its power is generated by renewable sources, nearly all of that (91 percent) from hydroelectric dams. In that sense, BC is a sneak preview of future power grids in jurisdictions around the world that are pursuing ambitious renewable-energy targets for 2030 and beyond. Electricity, however, currently represents only 18 percent of the province's overall energy use. Oil and gas still supply 66 percent of BC's fuel, and energy demand continues to grow. That's still a lot of the province's emissions to be addressed. And the best way to address them, as per James's tattoo, is to electrify as much of that energy use as possible and to expand the grid mightily to prepare it for fuelling cars and trucks and tractors, providing power to industrial processes, and all the rest.

Beyond heating and cooling buildings, the most important target for the project of electrifying everything—in British Columbia and beyond—is transportation. Worldwide, transportation generates 14 percent of all emissions. And in places where the grids have gone greenest, it quickly emerges as the biggest slice of remaining emissions. It's also the one that's most readily addressed from a technological standpoint—swap out internal combustion engines for electric motors in all the cars and trucks on the roads, radically expand transit and electric train networks, encourage "active transport" (bikes, scooters, feet) in urban spaces. But it's one of the thorniest targets from the inertial point of view of how people live their lives, what they're willing to change, and which infrastructure projects and policies they'll reward politicians for pursuing. Recall what I said earlier about how traffic and parking can, with staggering ease, cause friction at a civic meeting—nothing is as effective at making otherwise reasonable people act like panicked lunatics. Recall as well why I'm so impressed with Tesla's track record—not because I love a cool car but because I love a climate solution that sells itself on the simple merits of doing the job better. This is the case with electrified and other non-emitting transportation methods as well, but it's much harder to sell those to everyone than it is to sell one person a new Tesla.

Consider another jurisdiction that begins this next phase of climate solutions with an impressive head start: Costa Rica. There might be no place on earth that has embraced the "green" idea as fully as Costa Rica has. It was one of the first tourist destinations to brand itself as an "ecotourism" hub, and it has punched well above its weight in international conservation treaties and climate negotiations (Costa Rican climate negotiator Christiana Figueres spent six years as chair of the UN climate talks, including the breakthrough Paris summit). So when Costa Rica began unveiling ambitious climate goals circa 2007, the prospects for reaching them and establishing global leadership—even as a country significantly poorer than most of the largest players and biggest emitters—were very strong indeed.

Greening the electricity grid wasn't much of a hassle at all. Virtually the entire country abounds in streams and rivers racing down steep green mountainsides, replenished regularly by one of the wettest climates on earth. By 2017 Costa Rica's national grid was 99 percent emissions-free, and it started making headlines for going months at a stretch with no emissions at all. The following year, a young and ambitious new government announced even bigger plans. Its thirty-eight-year-old president, Carlos Alvarado Quesada, vowed that Costa Rica would be "one of the first countries in the world . . . if not the first" to eliminate fossil fuels entirely.

The international media were quick to gush. "Costa Rica Is Moving Toward Carbon Neutrality Faster Than Any Other Country in the World," a Vox headline read. Another, in Britain's *Times*, blared, "Costa Rica Leads the Way in Tackling Climate Change." The *Christian Science Monitor* added, "Costa Rica's President Elect Promises Zero-Emissions Transport." There is a tendency in these breathless stories from afar, however, to conflate electricity use (which represents less than 40 percent of all the energy used in Costa Rica) with *total* energy use. So let me point out that the single largest emissions source in the country is transportation, and further note that 70 percent of Costa Ricans live in a vast urban sprawl in the country's high, relatively cool Central Valley, around the capital of San José. Having spent a good deal of time in Costa Rica—and much more than I would have liked trying to get around in San José—I'm confident in saying that the nation's transportation system is a *long* way from its emissions goal.

Costa Rica's climate issues are almost all urban. Solve the transport problem in the Central Valley and you've solved it for the overwhelming majority of Costa Ricans. Transportation is responsible for more than 60 percent of the country's emissions, and most of that traffic is into and out of and around the Central Valley. The status quo, when the government first sized it up circa 2005, was a deep and growing dependency on internal combustion engines. Costa Rica's mounting prosperity had meant hundreds of thousands of private cars and new roads and highways to drive them on. The bulk of the country's freight moved by truck. And the only "public" transit to speak of was a vast, multi-tiered network of private passenger bus services, which, to their credit, provide transportation to and from even the remotest villages in the hills at very affordable rates, even as they burn diesel fuel in great volumes.

The most obvious way to address the relentless snarl of oil-burning traffic all those cars, trucks and buses had created in the Central Valley was mass transit. At the time, though, there was neither budget nor appetite for laying new track for dedicated commuter rail. Instead, after much more time and fuss than might seem necessary—roughly twelve years from concept to launch—the mostly moribund national passenger rail company, Incofer, set up a (very) modest commuter train network running along existing heavy-gauge track out of the old stations in downtown San José, lovely old buildings that used to greet passengers heading to and from the Atlantic and Pacific coasts. One line runs west to Alajuela, near (but not all the way to) San José's international airport, another east to Cartago, and a third into the suburbs south of the city. They burn diesel, and though the trains are often crowded at peak commuting times, the network is nowhere near frequent or extensive enough to make the slightest dent in the country's transport emissions. Costa Rica ignored the most basic tenet of a successful energy transition—it didn't electrify anything.

The next ten years and more will by necessity be one of the great ages of infrastructure development, and it must be an electric age. The Central Valley of Costa Rica, like everywhere else, aches for electric transport. In the Costa Rican case, it may make more sense to simply switch its established, comprehensive bus network to electric vehicles, and to give those vehicles their own dedicated lanes amid the gridlock. These sustainable urban design innovations are already up and running with great success in other parts of the world. Chinese cities switched to electric buses with staggering speed—the city of Shenzhen, for example, converted its entire municipal fleet of more than 16,000 vehicles to electric buses in only five years. And dedicated bus lanes resembling mass rail-transit lines—that is, bus rapid transit (BRT)—are particularly common (and well used) in Latin America.

The extensive TransMilenio BRT system in Bogota, Colombia, is considered one of the great transit triumphs of the past twenty-five years, with a transformative impact on mobility and livability for millions of Colombians of modest means. Similar networks, if less extensive, have been implemented in Buenos Aires, Lima, Guatemala City, Léon, Monterrey, Guadalajara and a dozen others (I can report first-hand that the Guadalajara BRT runs as smoothly and speedily as a commuter rail line). Most of these aren't yet electrified, but they are much more easily adapted that way than many other urban transport systems. Bogota's TransMilenio, for example, is in the process of converting its fleet to all-electric buses, with the first of more than 400 electric vehicles arriving in late 2021.

The logo on those new TransMilenio buses is telling. They are manufactured by BYD, China's largest electric-vehicle manufacturer, which began as a maker of cellphone batteries in the 1990s and now sells more than 30,000 electric vehicles per month in China alone. Mass transit has been a booming business for the past ten years and promises to boom further still in the decades to come, as cities and countries pursuing ambitious climate goals and post-pandemic recovery amp up their electrification strategies.

Asia is the emerging epicentre of both the transit boom and the new electric age in general. Roughly half a million electric buses are now on the road in China—about a fifth of all the buses in the country—while the United States still counts its nationwide electric bus fleet in the hundreds. Nearly a hundred major mass transit systems have been built since the turn of the century, all of them running on electrified track, and the lion's share of them are in Asia. This includes Chinese cities by the dozen, but the growth is continent-wide. Delhi completed its first subway line in 2004, added more in 2006 and 2011, and has two more lines in the works. Bangkok and Kuala Lumpur have built elevated rail systems. Dubai opened its first LRT line in 2009. Ho Chi Minh City's first subway line is under construction.

In North America, at least ten new LRT lines opened in 2019 and another twenty-eight were under construction by then. New streetcar lines started carrying passengers in 2018 in Milwaukee, Oklahoma City and El Paso. Global mass transit ridership expanded roughly sixfold from 2012 to 2017 alone. Combine this with the rapid expansion of high-speed rail in China, Japan, Korea, Germany, France, Italy and Spain—a fourfold growth overall in passenger-kilometres travelled—and you can see the electric age's steel-rail backbone beginning to take global shape.

That backbone, though, will not be enough to carry the whole world's transport demands. As much as urbanists and bike advocates and Chinese engineers and plain old boisterous boosters of high-speed trains and LRTs and bike lanes like me might want it to be otherwise, the planet's vast network of roads isn't shrinking any time soon. The automobile will be a major piece of the world's low-carbon transportation future. Or, as Bloomberg columnist Noah Smith put it in a recent piece, "We will not ban cars."

Smith's case verges on incontrovertible. He spent several years living in Japan and so uses it as his case in point. That country provides a powerful example as the world's leader in rail travel by a wide margin. Urbanists use "modal share"—the portion of travellers using a given mode of transport as their primary daily conveyance—as a shorthand stat to analyze transportation systems, and the share of Japanese transport that occurs by rail was 37.2 percent as of 2016. A little more than a third of Japanese residents, that is, use the country's comprehensive network of urban metros, commuter lines and bullet trains to meet their daily transport needs. (The nearest competitor, at 24.4 percent, is Russia. Compact, rail-covered European countries such as Germany, France and Italy score only in the single digits.) But despite being, as Smith puts it, "the most train-using country on planet Earth," more than half the Japanese population doesn't use trains as their primary mode of transport.

Japan, Smith argues, represents "an upper bound on America's ability to shift away from cars and toward trains." Even a concerted effort at urban densification and passenger rail expansion across North America, far beyond any seen in more than a century, would barely achieve Germany's more modest 8.6 percent modal share. Mass transit accounts for exactly one percent of all the miles travelled on land in the United States each year, while the motor vehicle's modal share across North America is 94 percent. The only parts of the world where it is less than 50 percent are South Asia and sub-Saharan Africa. A climate-driven rail expansion that succeeded beyond the most starry-eyed booster's most fevered dreams might see global modal share approaching 20 percent within a few decades.

In certain cities, of course, the market for rail, transit and active transport is already much, much higher, and the upper limit for commuter travel might well flirt with 100 percent. Cities such as Singapore, Hong Kong, Tokyo, New York, London, Paris and Berlin all have more transit users than car owners. In Copenhagen, as I've noted, cycling is the primary mode of transport for downtown dwellers, and the city's excellent transit system carries around another large share of people—only 26 percent of Copenhageners use a motor vehicle for their daily commute. But taking note of those cities merely leaves the overwhelming majority of the world to reckon with, where a motor vehicle of some sort is more likely than not to be involved in the day's travels.

All of which is to say that EVs are central to the global energy transition. This is not the greatest news for cities. They will have to continue fighting rearguard actions against the automobile's encroachment on all those urban neighbourhoods, building Better Blocks and pushing for more bike lanes even as they work to expand transit and train networks. But it's tolerable news for the transition in general, because EVs are more than ready to make the electrification of the transport sector a reality in the coming years. This is because the cost of batteries (which make up the greatest portion of an EV's price) is at or near the cost threshold of $100 per kilowatt-hour of battery storage, long believed to be the point at which electric cars can compete head-to-head on sticker price with gas-fuelled cars. Ford's glitzy rollout of its all-electric F-150 in the spring of 2021 was further confirmation that this threshold is very near.

Electrifying everything on the road is not, however, just a matter of cars and big trucks. Some of the earliest breakthroughs have come elsewhere—"last-mile" delivery vehicles for companies like FedEx, electric-assist bikes and scooters for short inner-city trips. Consider the case of India, the world's second-largest country and fourth-biggest greenhouse gas polluter. Of its 250 million vehicles registered to private owners, 190 million do not travel on four wheels—they are three-wheeled rickshaws and two-wheeled motorbikes and scooters. And so in India, the task of electrifying everything on the road has begun with the country's vast fleet of smaller vehicles. The government is aiming to switch as much as 80 percent of these to electric power by 2030, and there are already 1.5 million electric rickshaws on the road. That figure represents more than 80 percent of all the electric vehicles in the country (a figure that the often overly patriotic Indian press likes to point out is larger than China's EV fleet). At present the majority of these e-rickshaws are unlicensed, and often they are poorly made and powered by dirty lead-acid batteries. But entrepreneurial digital technology firms in India are already working to switch them to the industry-standard lithium-ion battery, and other companies have sprung up to lease or rent properly licensed vehicles to rickshaw-wallahs at terms designed to work with their low incomes and lack of capital. Whether that 2025 target is reached or not, India is a reasonable bet to be home to the world's first fully electrified three-wheeled transport system.

In the energy transition's innovative spirit, this emerging world of EVs is not just less bad, not simply a deletion of tailpipes and emissions. They actually do the job much better, with applications that will transform daily life in ways we've only begun to imagine. Consider a clever little household product currently being developed by a Montreal company called dcbel. The company's first "smart home appliance" will allow EV owners to easily integrate their vehicle's power needs with household energy production and use. It also allows the car's battery to supply power to the house—a feature that has made blackout-prone California the first market for the device. Dcbel has also begun a marketing push into Texas in the wake of its catastrophic 2021 blackout. When Ford debuted its much ballyhooed all-electric F-150, it similarly touted the truck's ability to supply a full three days of power to the average home. The electric car is not just a car but a workstation, a backup generator, an integrated home appliance.

There will surely be other features and opportunities like these once grids become less centralized and household power generation more commonplace. The Bornholm vision of the twenty-first-century value proposition—the home as power plant, the vehicle as power broker, the smart grid as innovation hub—will spread with each new technology like the one dcbel is developing. And that brighter future I first spied on the Baltic Sea's horizon is coming into view almost anywhere that a car sits in a garage overnight.

As an Aside: Small EVs Are Beautiful

Tesla sedans and Ford trucks earn the most hype in the electric vehicle world, but some of the most exciting EV developments are occurring at much smaller scale. In the first months of 2021, for example, the world's bestselling EV was the Wuling Hongguang Mini EV, a tiny two-seater made for the Chinese market by a manufacturing partnership that includes General Motors. Smaller EVs have also become popular in the shipping business. Amazon, FedEx and UPS have all made major orders of small electric vans to deliver packages from regional warehouses to their customers, and the courier service DHL has already converted 20 percent of its fleet to EVs. FedEx has pledged to go fully electric for its last-mile delivery fleet by 2040.

One of the biggest electric transportation success stories of the pandemic, meanwhile, has been a global boom in the sales of electric-assisted bicycles, or e-bikes. As of July 2021, e-bike sales were up 240 percent in the United States compared to the previous year, and countries around the world reported similar figures. In Europe, approximately 3.7 million e-bikes are being sold each year, a total expected to grow to 17 million by 2030.

These are still early days for the EV industry, but there's already a wide enough range of options emerging to suggest a significant swath of the world's all-electric travel will involve a vehicle much smaller than a pickup truck.

3.14 The New Industrial Age

The global energy transition is not now and never will be primarily a small-scale enterprise. It won't be local or artisanal. There might be small-scale applications such as a household solar array or a community-scale energy project, but to feed, clothe, house and move a planet of eight billion, the transition's solutions are by necessity industrial in scale, and they will have to cope with the problems of that scale. Solar got cheap at industrial scale. Electric vehicles will be manufactured by international auto companies that manage global supply chains. Electrifying everything will require a colossal expansion of grids the world over to supply power to all those EVs and heat pumps. And as the transition expands beyond those vanguard sectors, it will need industrial-scale solutions to industrial manufacturing problems.

Fortunately, solutions to some of the thorniest emissions problems are now at hand. Steel production, for example, is the source of about 7 percent of the world's emissions. Burning coking coal to generate the ferocious heat needed to smelt iron ore and limestone into steel is as old as the industry itself, but emissions-free processes are already being tested at laboratory scale. An entrepreneurial firm connected to the Massachusetts Institute of Technology (MIT) and backed by Bill Gates's Breakthrough Ventures, for example, is developing an all-electric steel production process. Other initiatives are investigating the use of hydrogen or biochar—a type of hot-burning charcoal made from agricultural and forestry waste—as fuel. Another start-up company backed by Gates's money, Heliogen, has developed a concentrated solar power system capable of generating heat of 1,000°C, with potential use in the production of glass and cement as well as steel. And the world's three largest steel-makers—ArcelorMittal in Europe, Baowu in China, and Nippon Steel in Japan—have all made pledges to reach net zero emissions by 2050.

The cement industry in general awakened late to its significant role in the climate crisis—cement production generates at least 7 percent of global emissions—but a range of new technologies has emerged to devise ways to shrink its footprint. One, developed by a Canadian company called CarbonCure, injects some of the carbon dioxide produced by cement-making into the product itself. CarbonCure first made headlines when it received significant funding from Gates's Breakthrough Ventures. Then, in 2021, the company was named co-winner of the Carbon Xprize, a $20-million award sponsored by Houston-based power plant operator NRG and the Canadian Oil Sands Innovation Alliance, after a multi-year competition for companies with the best new technologies for removing carbon from industrial processes. The other Carbon Xprize recipient, CarbonBuilt, has also developed technology to shrink the emissions from cement production, boasting an overall emissions cut of as much as 50 percent by using less cement and more recycled materials.

In the Saguenay region of Quebec, meanwhile, Elysis, a joint venture of aluminum producer Alcoa and mining giant Rio Tinto, has been launched to produce the world's first certified emissions-free aluminum. Producing the metal has been part of the economic backbone in that part of Quebec for generations. But recently Apple came calling, looking for ways to make good on its promise to eliminate the emissions from every material used in its devices. Quebec's grid, fed almost exclusively by hydroelectricity, is among the cleanest on earth, so it became the host for Rio Tinto's new aluminum production process. "This could be the biggest advance in aluminum production in 130 years," the

Washington Post raved. Hard hats soon adorned local and national politicians alike, and ceremonial shovels turned soil in the ritual so beloved by officials seeking to attach their names to impressive job-creation statistics. The resulting photo op makes a tidy shorthand case for climate solutions as an economic opportunity, in a way that even the least enlightened of politicians can readily understand.

There is no economic sector or jurisdiction untouched by the global energy transition. And if, for the first decade or so, some did their best not to notice the encroachment, now virtually all of them are awake to the shift. It's now routine, for example, even in the boardrooms of oil-and-gas companies (and especially in the boardrooms of their investors), to talk about an emerging global peak in oil demand on the horizon. Some forecasts see it as soon as 2025, and there was heated talk in the early days of the pandemic suggesting peak demand had just occurred during the last busy oil-burning weeks before COVID shut everything down. Other forecasts suggest the early 2030s, while the most intransigent corners of the oil industry prefer to convince themselves that peak demand won't happen before 2040.

Coal use has already begun to plateau globally and is in steep decline throughout North America and Western Europe. Even in Asia, the primary source of much of the world's demand for the past ten years, the trend is beginning to shift away from reliance on coal, as renewables have started winning the argument on cost alone. India, for example, is now expanding its coal-fired power plant fleet by 86 percent less than had been planned as recently as 2019. The Philippines has declared a moratorium on new coal plants. Overall, about 80 percent of previously planned coal plants across Asia had been shelved by the spring of 2021, largely because of cheap solar power.

The sale of passenger vehicles, meanwhile, peaked globally in 2017. Fossil fuels—oil and natural gas in particular—will continue to play significant roles in power production in parts of the world for at least another generation. But I'd wager that won't be the case in many places beyond 2050, and that their decline will begin sooner and happen much faster than current predictions indicate. Ten years ago, for example, the coal sectors in Europe and the United States did not see their own slide into obsolescence—now inevitable—looming on the horizon.

Big business in general, meanwhile, has proven far more sincere and thorough in its climate actions than many would have predicted. Perhaps no climate plan met with more skepticism than that of the world's largest retailer, Walmart, in the wake of its first sustainability pledge, in 2005. But by 2015 veteran sustainability journalist Marc Gunther surveyed Walmart's progress for *The Guardian* and concluded that "no fair-minded person . . . could fail to be impressed with the company's accomplishments." It had exceeded its first, modest emissions-reduction goal, doubled the efficiency of its trucking fleet, and ensured the preservation of a million acres of wildlife habitat. The company's overall emissions peaked in 2012 and then began a slow decline.

Walmart has been working in good faith with sustainable business leaders such as Patagonia and Seventh Generation, as well as the Rocky Mountain Institute, Amory Lovins's pioneering cleantech think tank. The Environmental Defence Fund has been its enthusiastic partner throughout. And Walmart, with its massive influence as the world's largest retailer, has accelerated efforts to reduce packaging waste and has given organic food a huge push into the mainstream. In 2015 the company launched "Project Gigaton," vowing to remove a full gigatonne of annual emissions from its supply chain by 2030. It has pledged to draw all its power from renewable sources by 2035, operate an entirely electric trans-port fleet by 2040, and eventually to "become a regenerative com-pany," with a net positive ecological impact on the planet. But yes, because this is still Walmart, after all, it has not slowed its com-mitment to rampant consumerism in the slightest. It will still sell you a gas-powered leaf blower as readily as a set of organic cotton sheets, and it has every intention of continuing to sell you more of it all into 2030 and beyond.

Which brings me to the central question of the mass-consumer-grade, industrial-scale side of the climate-solutions beat. The question arises from the more idealistic precincts of the climate movement, which would like to see not just shrinking emissions but a shrunken corporate sector. (It's not uncommon to see END CAPITALISM placards at climate protests.) In these same precincts, net zero pledges from companies like Walmart tend to be dismissed as weightless greenwashing. But is it more plausible that Walmart will be substantially less destructive to the climate in 2030 or that there will simply be no more Walmarts? Will Walmart and other large consumer and industrial enterprises that are wedded to the status quo continue to meet enough of their tentative, sometimes reluctant pledges to be part of the march to net zero by 2030, or will none of them survive that long? My reading of some of the more strident calls for a war-footing economy and eco-socialist overthrow of the capitalist order sug-gests that many people dream of a future free of Walmart and big-box consumer capitalism in general. If the latter scenario strikes you as at all likely, I'd ask you to explain where the hun-dreds of millions of current big-box customers will be obtaining their daily necessities after 2030.

If, therefore, it seems far more likely that the world in 2030 will still have Walmarts, how can they be left out of the search for solutions? The company is, after all, the single largest consumer-goods retailer on earth. In much the same way that no climate solution at a global scale can exclude, for example, India, there is no business side of that solution that doesn't include Walmart and Amazon and McDonald's and Unilever and Rio Tinto and 3M and Nippon Steel—and even Royal Dutch Shell and ExxonMobil and all their oil-and-gas brethren.

My point is not that the climate problem will be solved by these industries, but that in most aspects the problem now appears *solvable*. Even just ten years ago it was commonplace on the climate-solutions beat to hear the expert voices in many industries proclaim the inevitability of a continued status quo. The world needs steel, cement, glass, aluminum. Producing it all means burning coal and oil and natural gas (which is, after all, why you find steel mills near coal deposits and glass factories near natural gas wells). A decade later, the question hanging over most of these industries is not *if* they will change forever in response to the climate crisis but *when* and *how* they will change. That's real progress, and it will provide an enormous motive force to help accelerate the pace of change in the years to come.

3.15 Power and Justice

In May 2021, a company called Greenplanet Energy Analytics announced plans to build three new solar farms in southern Alberta. Together they would add about 68 megawatts of renewable energy to the province's power grid. The lead developer was a well-established clean-energy company called Concord Green Energy. So far, so commonplace—southern Alberta enjoys 300 sunny days per year on average, and the southeast corner of the province in particular has solar resources on par with Rio de Janeiro. What was so distinctive and welcome about the news was the other partner in this joint venture: the Athabasca Chipewyan First Nation (ACFN).

ACFN is a nation of Dene people in Fort Chipewyan, a small Indigenous community on the northwestern shore of Lake Athabasca, directly downriver from Alberta's main oil sands operations. In 2019 ACFN partnered with the other two First Nations in Fort Chipewyan—Mikisew Cree First Nation and Métis Local 125—to develop its first renewable energy project, the Three Nations Energy solar farm. Fort Chipewyan is accessible from the south only by air, river or (in winter) ice road. And like many communities in Canada's North, its isolation has long left it dependent for winter heat and power on generators fuelled by expensive diesel. The Three Nations solar array, which started generating power in early 2021, eliminated a quarter of the community's diesel demand. This means twenty-five fewer diesel trucks navigating the dangerous ice road each year, and a little more energy security for the community. It also means that ACFN is now in the clean energy business, and this latest announcement indicated it's happy to be there. ACFN has begun spreading its new-found expertise to other communities and generating revenue that doesn't depend on the mostly unwelcome oil sands operations. (Like many First Nations in the region, ACFN operates a lucrative oil-industry services company, considerable but still meagre compensation for the permanent shadow the industry has cast over its traditional lands.)

These new solar ventures are excellent news, and I'm happy to report that it goes beyond a one-off feel-good story in northern Alberta. By the latest count, Indigenous communities across

Canada were home to more than 2,000 renewable energy installations, and nearly 200 of those were medium to large-scale
projects like that first solar farm in Fort Chipewyan. The largest
share of these is in British Columbia, where a strong incentive
program—including grants of up to $500,000 each to enable
Indigenous participation in developing clean energy projects in
their communities—has been up and running since 2013, and
where Indigenous legal and land rights tend to be strongest
because many First Nations in the province are not bound to the
federal government by treaty.

ACFN's continuing story of follow-on employment and growth
is not unique either. Far from simply meeting their own needs,
numerous First Nations have been keen to enter the clean energy
business at a commercial scale. Several major wind farms in northern Ontario have been established by Indigenous-run developers—
Batchewana First Nation, for example, partnered in a 58-megawatt
wind farm near Sault Ste. Marie that opened in 2015. And in many
cases these projects have meant at least some local training, job
creation and knowledge transfer. In communities where energy
has long been expensive and unreliable, economic conditions harsh
and lacking in opportunities, and the general situation one of lingering colonial oppression, the emergence of an Indigenous-led
clean energy industry is a tiny ray of promise, and maybe even
equity, in a scene too often lacking in both.

In climate advocacy circles, it has become common to discuss
"climate justice"—the idea that climate solutions should address
social issues, such as poverty, inequality and racial disparity, as
well as the technical aspects of emissions reduction. The discussion is both tricky and necessary, but I've left it aside until now
because I'm not making a moral case for the global energy transition. I'm mostly making a practical one, highlighting the most

effective tools and their best applications to date. I'm talking about technologies, techniques, interventions and systemic changes that can operate at somewhere near universal scale—the scale of the energy regime that currently supplies fuel for everything from diesel generators in Fort Chipewyan to gas stations on the German Autobahn.

Because the climate crisis is planetary in scale, solving it comes with an obligation of universality that makes addressing social justice issues extremely complicated, to put it very mildly. There are no well-established universal fixes for income disparity or national prejudices or other social injustices. A solar panel manufactured in China can be installed to generate emissions-free power in southern Alberta just as readily as in southern China, but the same can't be said for civil rights legislation. The climate crisis will unquestionably take a greater toll on poorer communities and countries, and it will disproportionately punish the developing world and the global south, an intrinsic bias made even more unfair by the fact that the wealthy nations of Europe and North America spent more than a century getting rich by deepening the crisis. Compensating for this unequal distribution of benefits and consequences should be encouraged wherever possible. I would argue that the German taxpayers' enormous subsidy of the development of cheap solar panels is an example of how these inequalities have already begun to be addressed, however marginally and unintentionally.

I find it hard to argue from available evidence, though, that the global energy transition can also be a global fix for problems that have haunted human civilization for much (if not all) of its existence. Asking the purveyors of climate solutions to solve poverty in any systemic way is like asking the engineers of the Apollo project to end the Cold War. It is neither their area of expertise nor their jurisdiction. Trying to solve both at once means piling the fundamental reconstruction of global geopolitics and economics on top of possibly the single most complicated collective action problem humanity has ever faced. There is no addressing all of this with any one moral code or ethical system or transnational justice regime. No social movement in the history of democracy has ever taken on a project of anywhere near such scale. And the climate crisis will not wait for universal agreement among the world's communities and nations on right and wrong, when it comes to energy production or anything else. New technologies are by no means neutral, but they also aren't intrinsically limited to any one moral universe. A smartphone does not make a moral argument for its function as a tool, whether it's used to organize pro-democracy rallies or to spread fascist propaganda. The same is true of a solar panel.

Put another way, if there is a climate-justice argument to be made about Fort Chipewyan's solar energy projects, it's meaningless without cheap solar panels. And cheap solar panels rely on the operations of systems, industries and whole economies that are not now and may never be entirely fair, equitable, just and free. I might wish it were otherwise, but the climate crisis won't wait on satisfying that wish. The world will shift to a low-emissions, net zero economic order with or without fully addressing those greater goods. Or, more accurately, with *and* without.

Let's go back to Fort Chipewyan, as I first got to know it in the early fall of 2015. I was arriving by plane from Fort McMurray, where I was researching the oil sands business. Like many Canadians, I'd grown up knowing next to nothing about Indigenous people in Canada, and what little I did know was skewed to cast them in the harshest possible light. I grew up surrounded by institutionalized racism, is what I mean, and I'd surely absorbed a good measure of it. My primary personal contacts with Indigenous people were in the town of Grand Centre, Alberta, adjacent to the military base where I spent my early teenage years. The patrons of the tavern at the Grand Centre Hotel often spilled out onto the sidewalk outside. As my friends and I would pass on the way to the movie theatre around the corner, we were provided with a first-hand look at the devastating toll taken by abuse and alcoholism on the people of the Cold Lake First Nations. Of course, we didn't understand it in such empathetic terms. We were classic Canadian bigots that way at the age of fourteen. I couldn't have told you what cultural groups those people belonged to (Dene and Cree mostly), which treaty had absorbed them into Canada's colonial reserve system (Treaty No. 6, signed in 1876), what year they'd been granted the right to vote in Canada (1960), or how many of their children might have been all too recently abused in nearby residential schools.

As I flew to Fort Chipewyan thirty years later to report on the impact of the oil sands industry, I got to talking to my seatmate on the plane, and he turned out to be a Dene elder. When he learned why I was travelling to Fort Chipewyan, he offered to give me a tour of the community. I accepted and spent a day following him around town on foot, from the sites of long-gone forts and trading posts to significant local landmarks and viewpoints. We went to the small cemetery behind a former Hudson's Bay post and he told me stories about some of his people buried there. Then he invited me to a gathering of elders the next day.

I arrived at the grounds of the Athabasca Chipewyan First Nation elders' lodge to find a sort of community festival—elders were teaching Dene youth how to roast ducks and smoke fish, how to tan moose hides and carve wood. I offered a gift of tobacco in the manner my new acquaintance had instructed me, joined in filleting fish and hanging the fillets over a smoky campfire to dry, listened to the singing and drumming, and tried not to make a nuisance of myself. I comported myself well enough that the elders granted me permission to attend a meeting the next day with officials from Teck, the mining company that was applying to build a new oil sands mine, closer to the community than any other to date.

And that's how I learned how things mostly went for Indigenous people in the path of Canada's energy industry. The meeting was long and contentious. It began with a PowerPoint presentation on Teck's sustainability principles, but that was mostly aborted in favour of listening to the Dene elders explain why they felt none of their concerns had yet been heard. I'd hazard a guess that every Indigenous person in the room felt they still weren't. The meeting ended like an indifferent shrug.

The issues at stake in that meeting room—in that community, in this country—predate our awareness of the climate crisis by centuries. The power imbalances and injustices are nearly beyond calculation, let alone any adequate compensation and reconciliation. I can't tell you how to begin to untangle that criminal mess, but I do feel there's some truth in saying that the way the renewable energy industry has come to Fort Chipewyan is intrinsically less imbalanced and less predatory. There isn't much in the way of coercion. There is no pollution. The technologies are biased toward smaller scale and community engagement (a hallmark of the renewable power business in much of Europe, for example, is community investment and ownership), and there is something at least verging on an equal footing in the projects. These are tools that the three First Nations in Fort Chipewyan have found an opportunity to use, and now one of them has begun developing them beyond their own community, for their own profit and for the gain of other communities. That isn't nearly enough, of course, after all Canada has done to the Indigenous peoples of Fort Chipewyan and beyond. But it's not nothing, either.

3.16 Levers and Incentives

I've spoken already about the transformative fact of cheap solar power. It stands as one of the greatest achievements to date in the pursuit of climate solutions. Let's now revisit the story of solar's staggering growth to consider how important the right kinds of levers can be.

In the summer of 2008 I travelled to Silicon Valley to report on an emerging solar boom that was already inviting skepticism. Sure, the Germans were churning out solar panels at breakneck speed, and sunny California was basking in a little boomlet of its own. But solar power remained clouded in uncertainty. Without lavish subsidies, would it ever be affordable enough to compete with, you know, the *real* energy business? Wasn't demand already starting to plateau worldwide? Polysilicon PV wasn't even the hot topic anymore—all the industry action seemed to be shifting to thin-film solar cells made from exotic new materials like cadmium telluride and CIGS (stacked layers of copper, indium, gallium and selenium).

Silicon Valley is an odd place. Physically, I mean. Geographically. Its seeming urban centre, San Jose, is actually on its periphery, and the only other patches of urbanity are little villages of staggering wealth such as Palo Alto and Mountain View, tucked like luxury retail versions of Disneyland's Main Street, USA, into a formless mass of office parks and suburban strip malls. As a visitor, you can spend a lot of time driving up and down the chain-retail strip of El Camino Real, marking the distance by the recurring signs of fast-food outlets and wondering if you've missed the centre of gravity somehow. By some calculations, that centre of gravity sprawls along winding Sand

Hill Road, where many of the Valley's legendary venture capital firms have comfortable offices in leafy little office parks. I went to meet with a partner at one of those firms, Erik Straser, then at Mohr Davidow Ventures, which had recently moved beyond its usual digital technology investments to buy a stake in a next-generation clean energy company called Nanosolar.

Straser was not modest about the ambitions of the firm's new investments. "Sometimes I ask myself, 'If this company was successful, would people name libraries and public high schools after it?'" he told me. "Who made the steam engine? Who made the lightbulb? Who will those people be for the twenty-first century? Who's the person that made mass-market solar affordable?" The answer turned out (perhaps unsurprisingly) not to be a Silicon Valley venture capitalist. The Valley's interests shifted away from cleantech when the return on investments in next-generation renewables and hydrogen storage proved much less substantial and predictable than for new smartphone apps and social media companies. By 2012 Mohr Davidow was scaling back its cleantech portfolio, and Nanosolar shut down for good in 2013.

In the meantime, the real answer to Straser's question turned out to be rather mundane. Who made solar cheap? German electricity consumers and the Chinese government. Spurred by the explosive catalyst of Germany's feed-in tariff—steered through parliament by wily old Hermann Scheer—German companies filled their country with solar farms and rooftop PV arrays. For nearly a decade Germans paid premium prices for solar power, but hundreds of thousands of them also collected inflated fees for selling power to the grid from their own rooftops. The German approach, in the face of constant criticism from abroad but very little at home, made solar power mainstream. And then the Chinese government lavished even richer subsidies on its solar industry, and solar power got really cheap. And now it's becoming truly ubiquitous. More than 700 gigawatts of solar power had been installed worldwide by the end of 2020, nearly twenty times as much as the amount that existed in 2010. In certain places where the financing is especially favourable and the sunlight sufficiently ample, solar power is, according to the IEA, "the cheapest source of electricity in history."

The lesson here is one that would sound familiar to thinkers such as Joseph Heath of the University of Toronto (introduced in my earlier discussion of the limits of enlightened cooperation as a climate solution), who has spent much of the past decade arguing for the climate crisis to be treated less like an environmental catastrophe or corporate crime and more like the mammoth collective action problem that it is. Collective action problems, as Heath likes to point out, can most readily be overcome by new incentives. Charging consumers several times more for solar power than the going rate for other forms of electricity, as Germany did with its feed-in tariff, represented a massive shift in incentives for power production. It was seen as reckless, uneconomical, borderline heretical and potentially disastrous—and it made solar cheap.

I don't want to go too far into the specifics of government policy, because the needs and political options in any particular jurisdiction are so idiosyncratic, and also because one of the lessons of making solar cheap is that the momentum of the energy transition can overwhelm seemingly immovable policy obstacles. Texas, for example, is by far the US leader in installed wind-power capacity, not because the state government is such a fierce climate warrior but because there was so much money to be made as the cost of wind power plummeted. President Trump and his reckless gutting of American environmental regulations and asinine rhetoric about "beautiful clean coal," meanwhile, did nothing to either slow the decline of the coal business or hinder the growth of the renewable energy industry.

There have been other successful shifts in incentives over the past decade or so. In the United States, for example, "renewable portfolio standards," which mandate that a certain minimum of a state's power is drawn from renewable sources by a specific target date, have been the primary drivers of growth in the clean energy business. The state of California is among the global leaders in solar power, in significant measure because of its portfolio standard.

When I talked to Joseph Heath about overcoming the barriers to collective action on the climate crisis, though, what he really wanted to talk about was carbon pricing. Few incentives in a market economy are as effective as prices for changing collective behaviour in a hurry. When Heath gets going with his argument for correcting the market failure of greenhouse gas emissions by making polluters pay for creating them, his tone verges on awestruck. "If you look at which levers the government has that it can pull on the climate change file," he told me, "by far the most powerful lever is the pricing lever. A carbon tax changes the price of every single good in the entire economy. That is an unbelievably comprehensive and powerful intervention."

The merits of a carbon tax as the foundation of a collective response to climate change are well established. An economy-wide flat tax on every tonne of carbon dioxide emitted is simple and transparent. Setting up a carbon tax is fast and cheap, and it requires no elaborate bureaucracy to administer. And because fossil fuel flows are large in scale, readily counted, and pass through bottlenecks such as refineries and power plants, tracking carbon dioxide emissions is far easier than it is for most pollutants.

(Think of the difference between counting how much natural gas a power plant burns and how much phosphorus runs off any given farm to cause a particular algae bloom in Lake Erie.)

Of course, Heath is far from alone in his enthusiasm. There is agreement among economists and academics on the merits of carbon taxes to an extent rarely seen in the discipline. One carbon tax proposal in the United States was endorsed by 3,300 economists and other experts, led by former Federal Reserve chair Janet Yellen. In another study, 75 percent of environmental economists and others with climate policy expertise supported putting a price on carbon. This isn't just praise for an elegant theory. Where carbon taxes have been applied, in jurisdictions that cover roughly one-fifth of the world's emissions, they've delivered on their promise.

One of the most widely praised models is in British Columbia, which introduced an economy-wide carbon tax in 2008, one of the first jurisdictions to do so. In the first four years after it was enacted, per capita fuel consumption declined by one-fifth. Overall, the tax is credited with reducing emissions in BC to between 5 and 15 percent below business as usual, while also having an impact on the provincial economy that a 2015 study deemed "negligible." Which is to say that, contrary to the warnings repeated ad nauseam by conservative politicians, it did not destroy the province's economy. One of the most powerful aspects of BC's carbon tax is that, after it was up and running for awhile, it became not a topic for contentious debate but an almost invisible aspect of the operating system of daily life. In a 2018 poll, only 45 percent of the province's residents were even sure the province had a carbon tax. It simply wasn't an issue.

A 15 percent cut in emissions might not sound revolutionary— it might well seem barely evolutionary—but the carbon price is not meant to be a silver bullet. It's better understood as a vital part of the necessary recalibration of the whole economy, a way to place it firmly on the low-carbon track before beginning the more complicated work of wholesale decarbonization. It's a first step, and a strikingly simple one. Heath calls climate change "the worst public policy problem imaginable," and it's truly hard to dream up a worse confluence than the one between the scale of the looming catastrophe and the capacity of our political institutions to respond effectively. "But," he adds, "there's one point on which we caught an unbelievably lucky break. And that is that we can use a pricing solution, and that that pricing solution can be administered so efficiently."

It's a strong point, and easy to miss—a carbon tax is remarkably straightforward to operate. There's no clunky bureaucracy around carbon pricing. BC's government did not need to fill floors of a Victoria office building with accountants and analysts. Repainting the markings on the Coquihalla Highway each spring is likely a bigger draw on the province's resources than administering the carbon price. And perhaps most importantly, it represents a firm push on a true lever of power. Unlike a municipal government's symbolic climate-emergency declaration or even a non-binding national emissions target under the Paris agreement, the carbon tax is a genuine change in the baseline operating code of the economy.

What's more, carbon pricing scales up and out in ways that few climate policies have to date. California, Quebec and Ontario have already worked across the border in cap-and-trade carbon pricing regimes, as have the member nations of the European Union. China, wary of most economic interventions from beyond its borders, is experimenting with carbon pricing. And in a 2015 paper, Yale University economist William Nordhaus (who has since won the Nobel Prize for his work on climate economics) argued that carbon pricing could serve as a more effective basis for coordinated international action on climate change than the cumbersome UN treaty process.

It's worth recalling another of Joseph Heath's points, which is that cooperation between nations on the scale and scope that the climate crisis demand is not a well-established norm we're failing to use but rather a wholly unprecedented state we're still striving to reach. Precious few tools are already in place to facilitate that kind of collaboration, but it might be that carbon pricing is one of them. Consider another of William Nordhaus's ideas, the "climate club." A climate club, he explains in a 2015 article in the *New York Review of Books*, is a group of nations that come together under a common carbon price (he recommends a rigorously calculated "social cost of carbon," which might start around $50 per tonne). These nations would then trade freely with one another while imposing across-the-board tariffs on goods imported from nations outside the club, equal to the established carbon price. The club could also include a minimum income threshold to avoid penalizing developing countries.

As Nordhaus notes, the costs of acting on climate change are national but the benefits are global (in the form of reduced emissions). This leaves the issue "particularly susceptible to free-riding"—a type of problem in which the benefits of a collective action accrue equally to participants and non-participants alike. Even worse, climate change encounters the problem not just in terms of overall emissions today, but also in the form of "temporal free-riding" by the current generation, which accrues the benefits of fossil fuel use at the expense of future generations who will bear the bulk of the costs. (Here Nordhaus offers an implicit fist bump of solidarity with Greta Thunberg and her student strike movement, even if neither party might acknowledge it as such.) "A central feature of the club," Nordhaus writes,

> is that it creates a strategic situation that is the opposite of
> today's free-riding incentives. With a Climate Club, coun-
> tries acting in their self-interest will choose to enter the club
> and undertake high levels of emissions reductions because of
> the penalties for nonparticipation. . . . A Climate Club that
> ensures high prices of carbon emissions around the world, or
> the equivalent, is an essential step toward an effective policy
> to slow warming.

There are, to date, no established models to demonstrate the merits (or drawbacks) of this idea. But the first tentative steps toward a system resembling Nordhaus's climate club have been taken in Europe, where the European Union has been considering a "carbon border adjustment mechanism" as part of its ambitious post-pandemic climate plan. The EU already operates under a carbon pricing regime, a cap-and-trade system called the EU Emissions Trading System. The border adjustment mechanism would impose a tariff on all goods imported from jurisdictions without carbon prices, equal to the current price per tonne of emissions in their existing cap-and-trade market. That market had long been undervaluing the cost of a tonne of emissions—the price lingered below €10 from 2012 to 2018—but it has recently risen as high as €50 and is expected to increase even further as the border adjustment comes closer to reality.

In one modelling study of the EU's proposed climate club, a carbon price operating without a border tariff would result in as much as 25 percent of the EU's emissions "leaking" into other jurisdictions. This is clear evidence of the free-rider problem that the tariff is meant to overcome—producers of emissions-intensive goods like steel and cement would simply move their production outside the EU. In the model, the proposed border adjustment eliminated most of the leakage, especially from heavy industry. And it could even increase the overall reduction in emissions of all goods traded by as much as 5 percent, by encouraging trading partners to cut their emissions as well.

The encouraging takeaway here is that, like the physical tools of the energy transition, the policy mechanisms that can accelerate the drive toward an emissions-free economy are becoming much more effective. It's even possible that a decade or two from now, clunky Paris emissions targets will be a sidenote to the story of how carbon prices and climate clubs drove the quest for climate solutions. Then again, maybe not. After all, one clear lesson of the past ten years is that predictions are a fool's game. Another lesson is that the world is getting better faster than ever.

3.17 The Humming Twenties

In July 2019, the European Union brought in new regulations for electric vehicles, obliging manufacturers to equip them with an "acoustic vehicle alert system," or AVAS. At low speeds—the sort of speeds cars travel at when approaching intersections, for example—EVs will be required to make some sort of steady sound to warn bystanders of their approach. The AVAS corrects one of those good problems to have: the motors in electric vehicles are virtually noiseless, emitting not even a Jetsonian electronic whirr.

The first time I encountered an EV in the wild, the silence was one of its most striking details. This was in 2008, when I was in northern California reporting on the solar industry's first boom, and there were very few fully electric cars on the road. I was walking up a street in San Francisco's North Beach neighbourhood, and as I came to a corner I spotted a Tesla roadster parked just beyond the intersection. I'd never seen one in person, and I was gawking at it like I'd spotted a wild tiger on a safari when it pulled away from the curb with startling speed and whooshed past me down the street. I hadn't noticed there was a driver inside, and the car's famously quick acceleration transpired with less noise than a laptop booting up.

Automakers have not yet settled on a default sound for electric vehicles, but preliminary suggestions have involved variations on the theme of a distinctive purring hum. Will that sound become the aural signature of the decade ahead—a decade that could perhaps come to be known as the Humming Twenties?

There has been a renewed wave of interest in the Roaring Twenties of a century ago as the world emerges slowly from the grip of a global pandemic with obvious parallels to the "Spanish flu" pandemic of 1918 to 1920 and its legendarily carefree aftermath. And though the 1920s were as riven with conflict and strife as any other ten years of human history—ask how carefree those years were for the Black population of Tulsa, Oklahoma, or Armenians facing slaughter at the hands of the Turks—the decade did in some ways literally roar. It was an era of rapid industrialization and rising prosperity that saw the mainstream emergence of the automobile, the dawn of widespread electrification, and the introduction of time-saving household electrical appliances such as the refrigerator and clothes washer. Stock markets boomed, industry grew, and mass manufacturing hit full stride, resulting in rising wages and an unprecedented growth in the share of the population with leisure time. This new mass audience birthed a mass-market entertainment industry, enabled by the electrified technologies of the time—movies at the local theatre, jazz on the radio, live concerts and professional sports. In music halls and factories, on roads and commuter-train tracks, the years echoed with the roar of industry.

And now, emerging again from the devastation of a pandemic, with another era of technological advancement and reinvention on the horizon, might the decade of emissions reductions and electrifying everything hum like an electric car? I'd prefer a symbol more in harmony with the vibrant urban landscape that all the world's neighbourhoods should aspire to, but icons tend to emerge from novelty, and multi-unit infills alongside bike lanes likely lack the iconographic punch of a roadway full of whirring electric cars. In any case, this is my hope for the decade, and my reasonably safe bet on its trajectory: the years to come will hum as never before. Hum with the harmonies of less polluting vehicles, hum with heat pumps in place of gas-burning furnaces, hum with electrified LRT lines, hum with the electricity from solar-powered roofs feeding banks of batteries. In the first months of the pandemic, reports from cities around the world marvelled at the return of a more pleasing range of sounds—bird calls, the breeze in the trees, the human voice—to streets that had too long been enveloped in the nullifying roar of internal combustion engines by the score. The Humming Twenties will, I believe, righteously tip the balance further in the direction of quieter and more welcoming urban spaces.

As Vaclav Smil has explained, there have been three major energy transitions in humanity's history: the capture of fire, the shift to agriculture, and the rise of fossil fuels. The fourth, a wilful transition to clean energy, is now under way, and its continued expansion is inevitable everywhere in the world. Smil is, as I noted, a skeptic regarding its pace, arguing that it will take many generations, if not centuries. But then, the pace and promise of this global energy transition have been underestimated every step of the way so far. Among its great achievements has been to already erase, all but completely, the potential for the most catastrophic climate impacts that have been modelled by the IPCC and other scientific organizations in the past twenty years.

There's one scenario in particular that inspires rigorous debate and criticism on social media. Its shorthand among climate policy wonks is RCP8.5, a "representative concentration pathway" modelled by the IPCC in which emissions continue to rise almost unabated for the rest of this century, driven by a sevenfold increase in coal use, and resulting in 5°C of warming. It's the very worst case, and the source of the most apocalyptic climate-news headlines in the past twenty years. It has also been rendered essentially fictitious by the progress made toward the global energy transition in the past decade. Social media debates between climate scientists and energy policy wonks nowadays focus tenaciously around whether it's even responsible to continue to talk about it. Catastrophic outcomes are still very much in play, but a climate meltdown driven by five degrees of warming is not much more likely than the COVID pandemic turning into an actual zombie apocalypse.

The transition to date justifies optimism from virtually every angle. The past ten to fifteen years were merely a long, looping prelude of doing less bad—and it was still the start of the most rapid shift the world has ever seen from one energy basis to another. Clean energy has attracted $4 trillion in investment since 2004, the largest share of it ($1.75 trillion) in Asia. Wind and solar power are already the cheapest source of new power for two-thirds of the planet. In 2019, renewables provided 72 percent of all new electricity capacity added to the world's grids—and 54 percent of Asia's. Solar and wind grew by approximately 127 gigawatts and 111 gigawatts respectively in 2020 alone, bringing global capacity to 714 gigawatts of solar power and 733 gigawatts of wind—far beyond any forecast anywhere ten years before. This amounts to about four times as much renewable power in total as there was on earth in 2010, and more like fifty-seven times as much solar power.

Renewable energy was the only sector of the energy business that continued to grow throughout the pandemic. IKEA, the Swedish furniture company, is responsible for 1.7 gigawatts of renewable power all by itself. The Hoover Dam is, as I noted earlier, a 2-gigawatt power plant, and the stirring symbol of a previous era's engineering genius. IKEA is a Hoover Dam's worth of clean energy all by itself. Nearly two Hoover Dams' worth of offshore wind power (3.6 gigawatts) was installed in Europe in 2019. There are three Hoover Dams (6.4 gigawatts) of solar PV in Africa, a doubling in capacity since 2016. China has about 102 Hoover Dams of solar power alone, and the United States approximately 35. Germany, which is not particularly sunny, has 25 solar Hoover Dams. There are next-generation floating wind farms under construction off the coasts of Norway, Portugal, France and the United Kingdom. There is now more renewable power than fossil-fuelled power feeding grids in Europe. The state of Texas had less than a gigawatt of wind on its grid in 2013 and 40 gigawatts of coal; in 2019 wind overtook coal, 22 gigawatts to 21 gigawatts—eleven Hoover Dams of wind power scattered across the oil heartland of America.

The list of these transformations is potentially endless. There are 7 million electric vehicles now on the road, including about 500,000 buses. They've already eliminated at least a million barrels of oil per day from global demand, and they will hit the fat mainstream of vehicle sales, where Ford F-150s are sold by the thousands, only in 2022. At least 4,500 passive houses were providing shelter worldwide as of 2019. During the pandemic, active transportation—cycling and walking, mostly—increased by 42 percent worldwide (and by 162 percent in the UK). There are solar panels on the roofs of 1,000 Indian railway stations. In October 2020, for the first time ever, a renewable-energy developer, NextEra Energy, was the most valuable energy company in America by stock market valuation. The solar industry employs a quarter of a million Americans. The age is already humming with clean power and electric motors.

The 2020s will not hum, however, without considerable effort—lots of it, everywhere. The monumental task, for everyone tuned in to these new frequencies, is to will into being the much better world through their collective optimism and exuberance, to demand that it be made so. Recall the reflections of the European Climate Foundation's Laurence Tubiana, one of the architects of the Paris climate agreement in 2015, which I'm confident will come to look in retrospect like the world's first full, coordinated pivot in the direction of this much better world: "The . . . philosophy I worked from was that the real enforcement mechanism couldn't be a punishment or sanctions-based one: the enforcement mechanism was changing expectations. The belief that the transition is inevitable is the mechanism itself."

The belief that the transition is inevitable is the mechanism itself. And the motive force behind this mounting sense of inevitability does not emanate from any one source—not from the front ranks of student climate strikes, not from the corporate boardrooms where net zero pledges are made—nor does it always point in a single direction. It's both a protest movement *and* a slow recalibration of incremental political actions. It's a consumer's choice of new car *and* the realignment of multi-billion-dollar investment portfolios away from the mounting risk of fossil fuel projects.

It's corporate *and* grassroots, big business *and* local cooperatives. It's electoral politics, activism, civil society, corporate governance and retail consumerism. It isn't a single vector and it won't be tracked by a single metric. But it is under way, and it's accelerating, and it's enormous. It will explode across the Humming Twenties with a force far beyond what the world's seen to date.

To put a provisional date on the birth of the Humming Twenties, I would suggest the last two weeks of May 2021. Within ten busy days, a handful of news items sketched the contours of the new age. The first of these was the IEA's "roadmap" to net zero, an alarm call advocating for "unprecedented transformation" of the world's energy economy to keep global warming near 1.5°C and reach global net zero emissions targets by 2050.

A few days later, the Ford Motor Company unveiled its F-150 Lightning, the world's first all-electric pickup truck for the mass market. The F-150 is the bestselling passenger vehicle in North America. Although its debut was far from the first clear signal of the auto industry's abandonment of internal combustion engines— Volkswagen, BMW, Mercedes, Volvo, Nissan and General Motors have made commitments to massively expand their EV production—Ford's all-electric truck was the first clear guarantee that the shift would begin right away. The new F-150 would start rolling off assembly lines by mid-2022 at a price competitive with the best-selling conventional model. It's notable, as well, that Ford's launch made much of the electric F-150's ability to serve as a power source for heavy construction equipment, or even to supply a home with three days of power in an emergency. A brand built on the image of the no-nonsense, red-blooded American worker was pitching a significant swath of the value proposition I'd first spied on a remote Danish island to sell its flagship truck. In much the same way that, as I noted earlier, the net zero future will likely still have Walmarts in it, that same future might well have Ford pickup trucks as the leading progenitor of the vehicle-to-grid storage technology born at the very vanguard of the energy transition on Bornholm.

The following week, a court in the Netherlands ruled that by 2030, Royal Dutch Shell, the world's fourth-largest oil company, must cut its emissions to 45 percent below 2010 levels in order to comply with Dutch climate goals. Days later, at the annual general meeting of ExxonMobil, the world's largest oil company, two representatives of an activist shareholder group called Engine No. 1 were elected to the company's board—an unprecedented breach of the corporate barricades by avowed climate advocates. The Engine No. 1 reps were backed by major institutional shareholders, including BlackRock, Exxon's second-largest shareholder and the world's largest asset manager.

On their own, none of these developments was necessarily transformative. A few electric F-150s alongside hundreds of thousands of V8s; an aspirational, non-binding report from the IEA; a court ruling sure to spend years in appeals; two seats out of twelve on the board of a company that spends untold millions lobbying against climate action and for decades has led efforts to spread misinformation about the climate crisis. Taken together, though, as a ten-day snapshot of what the routine news cycle will look like in the years to come, they are a substantial signpost. There is no path back from here to a world where fossil fuels are dominant and clean energy an afterthought, no retreat from deepening action to slash emissions. The Humming Twenties have begun.

The large-scale pledges and plans that will shape this decade are increasingly in place. More than half the world's emissions are being produced in jurisdictions that are working within net zero frameworks, and the total is growing fast. China and the United

States, the two largest emissions sources, have both committed to net zero goals. China's near-term ambitions are focused on reducing emissions *intensity* (producing a smaller volume of greenhouse gases relative to the volume of economic activity), but its overall emissions are expected to peak by 2030, by which time more than half its grid will be powered by renewables. The sheer scale of that shift will mean hundreds of billions invested in the low-carbon economy every year. Of all the post-pandemic recovery plans, the European Union has the most ambitious on the climate front, while pacesetting Germany has deepened its commitment to the *Energiewende*, pledging to bring 14 million EVs onto its roads by 2030, to eliminate coal power around the same time, and to achieve net zero emissions by 2045.

The first years of the Humming Twenties are likely to see less dramatic results on the global emissions curve than advocates would like, but the longer-term picture improves by the day. In 2009, for example, the IEA estimated that global emissions in 2030 would exceed baseline levels by 44 percent; its current estimate is now 24 percent and steadily being revised downward. That's far from a crisis averted, but it's also far, far rosier than once projected. And it's a snapshot of an object in accelerating motion. I wish the trajectory was already pointing to an 80 percent reduction worldwide by 2030, but it is at least trending sharply downward. Let's see how much further the pursuit of a much better world can bend that curve.

The main engine of the Humming Twenties is cheap renewable power—far cheaper than most experts predicted, and growing cheaper still. A 2019 report by the French investment bank BNP Paribas asserted that, in the long term, "the economics of renewables are impossible for oil to compete with." In digital-tech circles there's a phenomenon called Moore's Law, which explains the wild explosion in computing power by showing that the number of transistors on a microchip doubles every two years. The clean-energy equivalent is sometimes referred to as an example of Wright's Law, a principle of innovation first articulated by aviation engineer Theodore Wright in a 1936 article on airplane costs, which states that for every doubling of the production of a technology, costs will decline at a fixed rate. One recent analysis found that the cost of wind power has been declining by 23 percent for every doubling in installed capacity, and the price of solar has dropped by at least 30 percent for each doubling. The rate for lithium-ion batteries is about 19 percent.

As a result of those plunging cost curves, the investments needed to keep the Twenties humming will be easier than ever to obtain. "Solar and wind are proliferating not because of moral do-gooders," a big-picture analysis by several Bloomberg reporters explained in 2019, "but because they're now the most profitable part of the power business in most of the world." In 2020, renewable energy expanded by more than 250 gigawatts worldwide, a new record. The IEA noted that analysts expect this scale of growth to become "the new normal" in the energy business. Rapidly declining costs have led to constantly revised estimates for the energy transition's quickening pace. A report by the consulting firm McKinsey projects that renewables will be the world's cheapest source of power "in most regions" by 2030. (Handicapping for the habitual underestimation of these technologies by mainstream analysts, I'd place money on that threshold being reached by 2025.)

This set of circumstances has decisively shifted the mainstream conversation around the energy transition from one of reluctance and skepticism to questions of who will capture the biggest opportunities from its rapidly expanding impact. New Jersey, for example, has taken to pitching itself as "the Houston of offshore wind," commissioning a major new port in the south of the state to help install the 16.5 gigawatts of offshore wind power that New York and New Jersey plan to build by 2035. California, America's largest state and its pacesetter in the energy transition, has set 100 percent as its statewide renewable power target, choosing 2045 for the end date and an interim goal of 60 percent by 2026.

America's tech sector, meanwhile, is pushing much sooner toward 100 percent renewable power for its operations. Apple has already hit that goal, while Microsoft intends to reach it by 2025, en route to pacesetting carbon-negative status by 2030. Amazon, Google and dozens of smaller firms are pursuing similar targets. The broader corporate world is following along, with net zero pledges and clean-power contracts fast replacing fuzzy sustainability commitments as essential accoutrements for any business hoping to remain in the good graces of consumers, attract top talent and stay competitive. A UN-sanctioned study in late 2020 estimated that, overall, companies responsible for combined revenues of greater than $11 trillion worldwide have committed to slashing emissions to zero. That represents an extraordinarily quick shift from vanguard to mainstream in the private sector—and a sturdy base from which to join the business world of the Humming Twenties.

The quickening pace is startling—and welcome. And the global default setting in the energy game has already begun to shift to renewables, as Vietnam has recently demonstrated. Vietnam mostly sat out the first quarter century of the energy transition, choosing the previous default of cheap coal for its power needs. When foreign banks began to shift away from funding new coal plants a few years ago, however, the Vietnamese government realized that solar power was as affordable as any competing power source. In two busy years, aided by a subsidy for small-scale rooftop panels, the number of solar installations in the country increased a hundredfold. In 2020 Vietnam was third in the world in newly installed solar capacity, with around 9 gigawatts added to its grid—four Hoover Dams in one year—trailing only China and the United States.

There is, in short, a great gathering wave building toward clean energy and shrinking emissions, emerging from nearly every jurisdiction and business sector. The present commitments are, to be sure, nowhere near enough. But the sense of inevitability is powerfully, euphorically transformative. Given that virtually every estimate of the energy transition's pace and scope for the past decade has been exceeded, and that the transition is now rapidly accelerating in pace, it stands to reason—doesn't it?—that the myriad pledges for 2025 and 2030 will feed an even more powerful wave of change throughout the Humming Twenties, beating targets and inspiring even stronger plans as it builds.

In the spring of 2021, the government of Colorado introduced a suite of new building code rules intended to push the state toward net zero construction statewide, just as British Columbia's pioneering Energy Step Code is doing in Canada. As these approaches proliferate from jurisdiction to jurisdiction, the Humming Twenties will surely witness a powerful wave of retrofitting, creating billions in new revenue for the construction trades even as it makes hyperefficiency the default setting for new and old buildings alike. By one estimate, BC's Step Code will mean more than $3 billion in additional revenue for suppliers of building materials in that province alone.

Electric vehicles, as I've noted, will become commonplace for the first time in the Humming Twenties. It's impossible to tell how many of the big automakers are truly euphoric about EVs and how many are merely keeping up with Tesla and the rest of the pacesetters, but in any case, the sharp shift away from internal combustion engines is well under way and will be an inevitability before the decade is through. Mercedes has said it will launch ten EV models by the end of 2022. Nissan plans eight models for 2023, with a goal of selling a million EVs per year thereafter. BMW expects that by 2025 as much as a quarter of its sales will come from its EV line, and Volkswagen intends to sell 1.5 million EVs in that year alone. GM, Toyota and Hyundai all expect to have more than twenty EV models in showrooms by the same year. The sale of new gasoline- and diesel-burning cars will be illegal in Britain as of 2030 and in California by 2035, and as EV sales boom and climate pressures mount, there will soon be many more such bans. The end of the hundred-year reign of the internal combustion engine on the world's roads will be essentially guaranteed by the end of the decade.

Not that long ago, the climate crisis wasn't even on the radar in many heavy industrial sectors. I was invited to speak at a structural steel industry conference in Winnipeg in 2009, and my role was essentially to walk the room through the very basics of climate change and its possible repercussions for builders of large-scale infrastructure. I felt as if I was confirming a rumour for much of the audience and talking about a sci-fi fantasy I'd invented for the rest. Barely ten years later, the pursuit of emissions-free steel, cement and aluminum is turning into a race to stay competitive, and I expect there will be a fight over the mainstream of those and many other sectors by the end of the Humming Twenties.

Already Volvo has launched a partnership with a Swedish steel-maker that is planning to use hydrogen generated by renewable power to produce steel, and a number of countries—including major industrial nations such as Germany, the UK, India and Canada—have begun plans for a partnership to fund the development of low-carbon steel and cement. India's largest steel producer, Tata, has meanwhile started work on its own pilot process for emissions-free steel, and Bill Gates has backed a new venture that plans to use concentrated solar power to manufacture both steel and cement. And trade groups representing 90 percent of the world's merchant shipping industry have begun lobbying the UN to put a price on carbon in their industry.

The margins of what was seen as possible in tackling the crisis when I first started on the climate-solutions beat are now the mainstream everywhere, from the energy sector to corporate boardrooms to the halls of government around the world. Taken together, they constitute all the elements of a global Green Marshall Plan capable of building a global emissions-free economy. The years to come will be transformative—and also undoubtedly combative, frustrating, chaotic, and in some cases (and in some places) catastrophic. But the general trajectory—toward much lower emissions, moving faster by the day—is no longer in question. It's a certainty. The energy transition will make its greatest strides yet, and it will make the world better in all the ways I've observed and in many I can't even imagine yet.

It won't be enough, of course—nowhere near enough. For that, the Humming Twenties will also have to be a decade when certain kinds of science fiction are turned into reality.

3.18 Adventures in Necessary Sci-Fi

In 2021 the Canadian Institute for Climate Choices, a climate policy think tank, modelled dozens of complex future-tense scenarios to determine whether Canada could achieve its net zero target by 2050. One of its report's main findings was enormously encouraging. About two-thirds of the emissions cuts necessary to meet Canada's 2030 climate targets and put the country on a path to net zero, the Institute found, can be accomplished by using existing technologies and techniques in their current state. Even without significant new investment and innovation, in other words, the bulk of the work can be done with the existing tools, and there was a range of options for completing the job by 2050. The report concluded that net zero is an achievable goal from a technical standpoint. This is excellent news—scientifically rigorous confirmation that the climate crisis can be fully contained.

The report's other main finding is more troubling. The remaining one-third of reductions by 2030, as well as a significant chunk of further cuts needed to hit net zero, would require what the report called "wild card" solutions: "high-risk, high-reward solutions still in the early stages of development." Some of these climate solutions are already at the field-testing or limited commercial development stage, but others exist only as pilot projects or lab experiments. Overall these technologies lean toward the outer limits of what seems viable at the scale, speed of deployment and affordability the climate crisis requires, especially compared to the much better world of cheap solar panels and electric cars already emerging. They fall into a wide range of technical and industrial categories— everything from agricultural techniques and planting trees to advanced nuclear technology and high-tech "carbon removal." There's no easy catch-all term for this stuff, so I'll try to honour both the plausibility and the experimental terrain by calling it "necessary sci-fi"—not fiction made real, but futuristic technology that must become everyday reality in a time horizon shorter than the production schedule of James Cameron's *Avatar* sequels.

One analysis after another has reached conclusions similar to the Institute's net zero study. A low-emissions future—even a no-emissions or negative-emissions one—is technically feasible, and a huge chunk of the work can be done by existing technology. But there's inevitably a final wedge of emissions, vital to address, that will be very hard to eliminate without some pretty spectacular technological innovations between now and 2050. The UN's IPCC has modelled more than a thousand future-tense scenarios to examine how a wide range of emissions profiles will affect the earth's collective future. There were 116 of those scenarios in which warming was limited to less than 2°C this century, and 101 of them relied on "negative emissions" technologies, none of which yet exist at commercial cost and scale, to achieve the goal.

Michael Liebreich, founder of Bloomberg New Energy Finance (BNEF)—the analysis team that has probably come closest to predicting the trajectory of the global energy transition to date—describes the situation in terms of what he calls the "Three Thirds World." The present path, he explains, will by 2040 create a world in which one-third of all electricity is generated by wind and solar, one-third of all the vehicles on the road will be electric, and efficiency efforts will mean that one-third more GDP is produced from every unit of energy. Liebreich describes this three-thirds trajectory as "extraordinary," exceeding the expectations he brought to the founding of BNEF in 2004. The problem is that, to truly tackle the climate crisis, the world must stretch far beyond those three thirds. "It will not be enough to reach the goals of the Paris Agreement," he wrote in a 2018 post. "There is no emerging new orthodoxy on how to decarbonize the rest of the economy."

Liebreich, for his part, has joined the board of directors of one of those boundary-pushing technology firms—one that just happens to be headquartered barely fifteen minutes by bike from my home in Calgary. The company is called Eavor Technologies, and it claims to have uncovered a cost-effective way to unlock one of the most ubiquitous and yet elusive energy sources on earth—geothermal energy. Human beings have used geothermal energy, of course, for as long as we've been visiting hot springs. In places where it is especially abundant and readily harnessed—the volcanic island nation of Iceland, for example— it is a mainstream energy source. Geothermal heat pumps of various sorts have been used for home heating and other purposes for years, but those systems tend to be cumbersome and more expensive than conventional sources of heat. Eavor's trick is to harness the free and limitless propulsive force of hot and cold liquids to create a closed, constantly circulating loop of multiple wells, in which cold water runs underground down the pipes, gets warmed by the earth's stored heat, and flows back up to the surface through a heat exchanger, for use in heating or power generation. The water, drained of its heat, naturally flows back down the pipe, pushing warmer water up and around. The minimal energy input required is how Eavor claims to have cracked geothermal's cost problem, giving it, Liebreich says, "the potential to change the world."

Over-torqued hype is a hallmark of the green frontier, of course. In my years on the climate-solutions beat, I've watched groundbreaking technologies like hydrogen-fuelled cars and monolithic concentrating solar towers fail to make it far from the test site, while tried-and-true approaches such as building larger wind turbines and making buildings more efficient have done the bulk of the work propelling the energy transition. That said, Liebreich did not build his career chasing hype, and Eavor is far from alone in its efforts. Companies around the world are receiving significant investments to try to vault geothermal energy into the mainstream and transform it into an affordable, potentially ubiquitous source of heating and cooling that could very well become a major piece of the energy transition. In fact, geothermal is the focus of one of the first spinoffs from X Development, the "moonshot factory" set up by Google in 2010 to generate necessary sci-fi solutions for thorny problems like climate change.

In addition to the self-driving cars for which it's best known, the X lab has investigated numerous potential climate solutions, including the use of molten salt for energy storage and the generation of clean fuel from seawater. If an X project isn't "moonshot" enough—that is, if it's viable right away—it gets spun off into its own company. That's how it went for Dandelion, a start-up created in 2017 to commercialize a smarter, cheaper, household-scale geothermal heat pump system. The Dandelion Air system has been installed in only a few hundred homes to date, but in 2021 it received follow-on funding from Bill Gates's Breakthrough Energy Ventures. The company has developed robot prototypes to make drilling boreholes for geothermal energy more efficient and aims to expand quickly to 10,000 installations per year. This is still far from the ubiquity of the PV solar panel, but it's worth recalling that the first PV cells were developed by a telecommunications company's experimental wing, Bell Labs, and that only twenty years ago, putting those panels on 100,000 German roofs was considered an ambitious jumpstart for the industry. As the energy transition accelerates, there is every reason to believe the spinoffs of blue-sky think tanks such as Google's moonshot factory will yield mainstream solutions. Turning seawater into fuel, for example, turned out to be technically viable—it was just too expensive at the time, and so it's back on the shelf for now.

This is the terrain of the green frontier. There are amazing discoveries, stuff that will make your eyes goggle like a Looney Tunes character, and there is hype—and often there is real potential behind it. And nobody really knows what will work. Back in 2008, for example, I visited an industrial park in Canberra, Australia, to investigate "solar dye," a substance almost like paint, generating power. The dye was made of nanoparticles called perovskites, and there are labs still working on turning it into solar paint and investors salivating at the notion of coating every roof in the land with the stuff.

Several governments, including Canada's and China's, and a number of deep-pocketed companies around the world are vying to make the next-generation nuclear power plants known as SMRs (small modular reactors) a big player in the energy transition. The technology is viable, and the plants are smaller, cheaper and less wasteful, as well as basically immune to meltdowns. They could be a crucial piece of the puzzle. In 2018 the Canadian government invested $50 million in a Vancouver-area company called General Fusion, which is developing a commercial nuclear fusion technology. In the first months of 2021, General Fusion launched a partnership with Britain's nuclear energy agency to get a commercial-scale model of its plasma-driven reactor up and running by 2025. Fusion technology has long been a mainstay of sci-fi—it's the power source for Iron Man's suit in the Marvel movies—but the common insider's joke about fusion is that it has been just another twenty years away from providing cheap, ubiquitous clean energy for the last fifty years.

Any of these technologies could see rapid advancements, becoming as commonplace ten or twenty years from now as a Tesla sedan is today—or none of them could. The encouraging sign is that there is such frenetic effort, and readily accessible investment capital, being directed at the problem. This too is an asset that the transition did not have at the start of the previous decade.

In the meantime there is unprecedented interest—and money—being directed at some less sci-fi but no less promising stuff. Hydrogen fuel, for example, has become a central focus of the next wave of policy incentives and big investments as the world's governments plan their post-pandemic economic strategies. "Green hydrogen"—generated from renewable energy via electrolysis—is a prominent feature of the European Union's pandemic recovery plan, which includes a $550-billion hydrogen strategy. Hydrogen is emissions-free, efficient, and technically straightforward to generate, and the EU sees it as a crucial source of power for decarbonizing industrial sectors like steel and cement.

In Canada, meanwhile, governments at several levels are investing in "blue hydrogen," generated from natural gas. By transforming gas into clean-burning hydrogen and burying the carbon dioxide in the process, blue hydrogen's champions claim it can shrink emissions by 95 percent compared to using natural gas directly as a fuel. The Alberta and federal governments have partnered with a US company called Air Products on a large-scale, $1.3-billion blue hydrogen plant near Edmonton, which is intended to be up and running by 2024. Blue hydrogen generates much less enthusiasm among climate advocates than green hydrogen, because it perpetuates the economic viability of a fossil fuel. On the other hand—in the same way the low-carbon world will still have Walmarts—natural gas remains a common fuel and the emissions reductions are real. Blue hydrogen represents an opportunity to ease a significant and still powerful sector of the fossil fuel business into the energy transition.

And hydrogen is only one of several low-emissions fuels drawing large shares of investment cash and enthusiasm. There is a whole wide world of alternative fuels—other promising technologies include biofuel from crops and agricultural waste, and biomass heating and power plants that burn straw and wood chips. And though each is freighted with its own troubling compromises, the odds are the world may need at least some of them, because they are still better than burning more fossil fuels.

Airplanes skipping from continent to continent on clean-burning hydrogen, big-rig trucks and mammoth supertankers powered by plant-derived biofuels, neighbourhoods heated by waste straw—these are all encouraging developments on the horizon. But none of it is enough. *All* of it is not enough. As I said, removing carbon dioxide directly from its industrial sources plays a significant role in 101 of the 116 IPCC scenarios in which the climate crisis is sufficiently addressed, which is near enough to all of them to describe a necessity. Fortunately, there's a fairly frenzied rush to do just that, ideally at sufficient scale to create a whole new trillion-dollar business sector called carbon removal. For a long time it was referred to as CCS—"carbon capture and storage"—but CCS is simply one category in a growing array of negative-emissions technologies.

The first wave of carbon removal has been in many ways underwhelming. As of 2019 there were nineteen large-scale CCS projects in operation worldwide, and together they had captured and stored 25 million tonnes of carbon dioxide. That amounts to about 3.4 percent of the total greenhouse gas emissions that year in Canada alone, which provides some sense of the chasm between current capabilities and the scale of the problem. By comparison, a partnership between the Dutch government and a number of energy companies intends to have a CCS project called Porthos up and running at the port of Rotterdam by 2024, where it is expected, out of the gate, to store 25 million tonnes of carbon dioxide per year under the North Sea. And Porthos is but one of a cluster of new CCS projects in and around the North Sea; others are in the works on the English and Scottish coasts.

Much of the buzz in recent years has shifted to a single technological holy grail: direct air capture, which is to say sucking carbon dioxide directly out of the atmosphere. In concept, this sounds the simplest. Never mind looking to engineering new kinds of smoke-stacks and filters and the rest; simply pull air into an apparatus that can separate the carbon dioxide from the other elements and bury it or turn it into something useful and emissions-free. In practice, however, the engineering required to isolate and extract 410 parts of carbon dioxide from a million parts of air is complicated, in the same way it might be to pull a few needles from a cloud of hay-stacks as they blow across a field in a raging storm.

The direct air capture business doesn't really amount to a nascent industry yet—there are only two companies on earth that have built up-and-running direct air capture facilities. One is a Swiss company called Climeworks, which has a model of its direct air capture technology running in Iceland. The Climeworks plant removes carbon dioxide from the emissions at a geothermal power plant and buries it in a rock formation below ground, where it hardens over two years into stone. The other, which has been swathed in the most hype, is a company called Carbon Engineering, which has its headquarters and a test facility in Squamish, BC, north of Vancouver. Carbon Engineering has attracted tens of millions of dollars in investment capital from Breakthrough Ventures, Thomvest (an investment firm run by one of the heirs to the Thomson Reuters fortune), and Chevron Technology Ventures and Oxy Low Carbon Ventures, both of which are investment arms of major oil companies. In 2019 the company entered a partner-ship with Occidental Petroleum (Oxy Low Carbon Ventures's parent company) to build its first commercial-scale direct air capture system in the oil fields of West Texas. It will be the largest direct air capture project in the world when it begins operations in 2023, pulling carbon dioxide from the dry Texas air and inject-ing it into depleted oil reservoirs, where it will aid in extraction of the last barrels of oil. Carbon removal in the interests of oil production, then. The optics are a little awkward, but the oil recovery will create enough revenue to make Occidental willing to sink capital into an experimental technology that won't make money on its own.

Carbon Engineering's long-term plan is to combine the carbon dioxide it captures with electrolyzed hydrogen to produce low-emissions synthetic fuel. The long-standing assumption in the direct air capture business is that $100 per tonne represents a minimum threshold for commercial viability. Carbon Engineering has published exhaustive research showing that at present its technology can do the job for between $94 and $232 per tonne. Not completely under the threshold, in other words, but close enough already to generate considerable euphoria, especially in parts of the conventional energy business that don't otherwise see particularly bright futures for themselves in a net zero economy.

There is no guarantee, however, that carbon removal of any sort will work on the scale and timeline needed. No guarantee, either, that renewable energy or hyper-efficiency or the long-sought dream of nuclear fusion will find a way to eliminate the significant last wedge of emissions that must be eliminated to reach net zero and prevent the climate crisis from becoming a wider catastrophe. Which means the solutions beat must include at least a consideration of geoengineering—deliberate further alteration of the planet's atmosphere—using methods such as seeding the high atmosphere with reflective particles to reduce the damage caused by global warming.

"If people are serious about the two-degree limit," British journalist Oliver Morton writes in *The Planet Remade*, his thoughtful and meticulously researched 2016 overview of the geoengineering field, "one or another form of geoengineering needs to be treated as a real possibility. The risks, costs, politics and practicalities need to be debated in a process which admits that there is more to climate change than emissions reduction and adaptation." As Morton notes, not only do the technologies exist, some are already in common use. As of 2013, more than forty countries worldwide had seeded clouds with various particles to suppress hail or increase precipitation, and the Chinese government notoriously used geoengineering to reduce cloud cover and smog for the 2008 Beijing Olympics. The possibilities verge on the truly sci-fi, and to some the topic is almost unthinkably dystopic. But then, so is the terrain as the world moves much further beyond 2°C of warming.

I won't handicap the various potential technologies—many are in their infancy, and Morton analyzes them from a much deeper knowledge base than mine. And I can't say with any kind of certainty whether the better world beyond 2030 will need small modular nuclear reactors or giant CO_2-sucking fans or a strato-spheric blanket of artificial cloud cover. What I can say is that I completely agree with Morton on the point that if you take the need to keep warming as near to 2°C as possible (if not below), then you cannot dismiss any technology out of hand. There will be some sci-fi in the mix, and it will be completely necessary.

4.0
THE AGE OF TRANSITION

When the pandemic forced the world indoors, it evidently encouraged a lot of binge-watching of older prestige TV series. There were so many memes and GIFs from *The Sopranos* bouncing around social media, for example, that it seemed like the show had just been released. And it reminded me of a central theme in that tour-de-force series (widely lauded as the show that launched the golden age of binge-worthy TV). In the very first episode of the first season, mob boss and patriarch Tony Soprano passes out from a panic attack at a family barbecue and winds up in psychotherapy. At his first session, his therapist asks him why he thinks he fainted. "Stress, maybe," Tony answers. He pauses, scratches the back of his neck and looks warily around the room. "The morning of the day I got sick, I'd been thinking it's good to be in something from the ground floor. I came too late for that, I know. But lately I'm getting the feeling that I came in at the end. The best is over."

This is a theme that recurs in many of the celebrated TV series that emerged in the wake of *The Sopranos*. Decline, exhaustion, the entropic spiral toward chaos and collapse. Walter White's futile and then brutal search for transcendence beyond a deadly illness and a dead-end life (*Breaking Bad*). The endless cycle of drugs, arrests, corruption and successive crime kingpins in the deathtrap the dealers on Baltimore's streets call simply "the Game" (*The Wire*). The bloody and ultimately futile quest for the Iron Throne while a much more apocalyptic threat mounts in the northern wastes (*Game of Thrones*). Every episode of *Six Feet Under* is set in a funeral home and structured around a death. The pointlessly cruel and corrupt prison world of *Orange Is the New Black* is self-perpetuating and inescapable. Everyone is trapped, all is decline. And there are always at least a couple of zombie apocalypse series up and running at any given time, in case you like your hellscapes a little more literal. Doom, as I said, is easy.

The show I found myself bingeing during the pandemic was more about generational transition than about things falling apart and centres that can't hold: *Mad Men*. The opening credits feature a silhouette of the show's male protagonist, Don Draper, tumbling from a building, but this is not a series about giving up or even falling apart. It's about change and loss and new beginnings. Don is not who he claims to be—he stole his identity from a man who died in a trench next to him in the Korean War—and the masquerade haunts him, but it's the inevitable shift in the culture around him that leaves Don lost and flailing. The same shift, though, provides a platform of escape and rebirth for the show's other protagonist, Peggy Olsen, who is also fleeing a past that felt like a prison. There is no final loss or ultimate gain. There is only transition.

In the third season of *Mad Men*, Don Draper meets Conrad Hilton, the hotel magnate, by chance at a wedding. They become friends of a sort, and Hilton teases Don with the promise of a lucrative deal to produce his hotel chain's advertising that never fully materializes. Along the way, they discuss guiding philosophies and the nature of the world as they understand it. Hilton invites Don to his suite at the Waldorf Astoria late one night to ruminate on the place of America in the modern world. "The Marshall Plan," he says to Don. "You remember that? Everyone who saw our ways wanted to be us." In a show that obsesses over the theme of masks and façades and sales pitches—the way people present themselves and what it can cost them—the line hangs heavily. America's confidence, its sense of inevitable triumph, was a powerful asset in the postwar world. Later in the episode, Don echoes Hilton's philosophy back to him in a pitch for his ad business. "How to lure the American traveller abroad," he says. "What more do we need than a picture of Athens to get our hearts racing? And yet the average American experiences a level of luxury that belongs only to kings in most of the world."

There is hubris in all this, of course, and imperial blindness. But there is also a core of truth—America did project its image of greatness to the whole world, and very few corners of that world were unchanged by its towering ambition. Some were trampled by it, many others drawn to it—as immigrants or imitators, as envious or contemptuous enemies. No one felt neutral about America in the generation after the Second World War. It was a powerfully enticing avatar of that sense that, as Tom Petty put it, there was *a little more to life somewhere else*. I've spent nearly twenty years on the trail of climate solutions, and I have seen no other social, economic or political force as powerful as that.

The global energy transition has to be not a flight from danger but a march, even a race, toward a better world. Its origins in climate politics, which in turn grew out of the environmental movement, have freighted it with the baggage of nostalgia that it must let go. There is always a backward gaze to environmentalism—it is at its core about recapturing a lost Eden, restoring a fallen world. And even if there's real truth in its gaze, this nostalgic aspect, coupled with its limits as a holistic worldview (as George Marshall has argued) and as a political program (as David Roberts has argued) mean that environmentalism is simply not a large enough force to compel the rapid, dramatic transformation needed to address the climate crisis.

The Green Marshall Plan—driven by dark euphoria, delivering the value proposition for a better life in the twenty-first century, open to necessary sci-fi—this is not the end of a trail but the start of an exciting new one. This is real motive force. And these aren't end times. This is a transition. Plausibly optimistic by necessity, rendered darkly euphoric by its context, with the trends all moving the right way, finally. It could even be a sort of golden age—if you want it to be, and if you make it so.

I thought about trying to piece together some future-tense snapshot of this better world in 2040 or 2050, but then it struck me that it would be foolish, and it would run counter to the spirit of these Humming Twenties just begun. I have no better idea, really, about how this will play out than anyone else, and the climate-solutions beat is littered with inaccurate forecasts and misguided predictions. More than that, the future is, as ever, unwritten. I've presented what I consider to be the best tools on offer to build a better world, but ultimately I'm not in charge of drafting the blueprints. Odds are that whatever tools are used to build it will be less radically different than anyone might predict right now, because the arc of history moves more slowly and methodically than many of us might sometimes want, even in the midst of a crisis. I doubt that world will be a jumpsuited sci-fi utopia, and it certainly won't be a global eco-village, but it won't be an apocalyptically scorched earth either.

The one thing I'll say by way of prediction: the future will be warmer. It looks exceedingly, almost impossibly difficult to imagine sufficient effort, even in the Humming Twenties, to slash emissions to a level where global warming stays below the 1.5°C target. But the target is not a scientific threshold. As Myles Allen, one of the lead authors of the IPCC's special report on 1.5°C, has written, "Bad stuff is already happening and every half a degree of warming matters, but the IPCC does not draw a 'planetary boundary' at 1.5°C beyond which lie climate dragons." So if not quite 1.5°C, then as little beyond that as possible. And the progress on shrinking emissions should only accelerate from year to year. If this were a Vegas betting pool and we could place money on an over/under line, I'd take the under on 2.2°C in a heartbeat. I'd also take the under on 2°C, even if right now it still looks like a bit of a dark horse. I'd even place a bet on, say, 1.7°C. All of which mean big trouble, but still far from the end of days. What that means in scientific terms is the 2.5°C scenarios now strike me as verging on the worst case. And the 2.5°C case is undoubtedly terrible in many ways, but it's another universe from the 4° or 5°C that feeds nightmares of planetary-scale dystopia. And to be honest, I think if the climate stabilizes anywhere below 2°C, it will come to be regarded as one of the most amazing collaborative projects humanity has ever accomplished.

What I mean is dammit this transition *has* to be optimistic. It has to have some excitement to it, at least a little exuberance, the promise of euphoria. People, masses of them, don't build something much better in panic and terror. Life rafts aren't sturdy, spiffy, stylish new boats; they are simply a necessity in an emergency. No one lives on a life raft—they are abandoned as soon as possible.

Let's instead build a much better world. Or try to—and in that effort make something at least much more durable and admirable. Something maybe even a little enviable. I have spent a lot of the past two decades in enviable places. Irresistible places. Sometimes even euphoric places. That much better world is waiting.

That sales pitch? That *little more to life somewhere else*? A level of luxury—by which I mean stable, secure, emissions-free, open to the future without fear—once known only to kings? That can win. Trust me, that can win.

ACKNOWLEDGEMENTS

It always takes what seems like at least a village or two to juggle the resources, working time and household duties to write a book, and this has turned out to be at least doubly true for writing one during a pandemic. Everyone's personal COVID story is, I'm sure, full of chaos, anxiety and heartbreak; for my household, with multiple immunocompromised members and a disabled child who hasn't been able to set foot in a physical school since March 2020, ours has been largely a story of constant improvisation amid mounting and waning waves of stress, during which we cobbled together multiple villages worth of friends, neighbours and family to help us manage childcare, paid work, the completion of my wife, Ashley Bristowe's, brilliant memoir, *My Own Blood*, and not dying.

I lack the space and the clear memory for a full accounting, but if you're reading this and you were one of the many who brought us a meal or stopped by for a sanity check or took my son to the river to throw rocks and buy me another hour to work on this book, know that I am deeply grateful. And that the most powerfully transformative aspect of this whole catastrophe for me has been the constant reminder that there is nothing ever built by human hands more important to our collective survival—in a pandemic, in the climate crisis, in any circumstance—than a strong community full of compassionate people.

Among the many crucial members of my community as I completed this book were: my father-in-law, Bruce Bristowe; my parents, John and Margo Turner, who endured the grind of full quarantine so that they could welcome their grandson to Nova Scotia for several months of pandemic caregiving; Alicia Carvajal, the best caregiver my son's ever known; Catherine Mercer, who not only provided a vital online lifeline to my son's therapy through the darkest months of the pandemic but has gone above and beyond her professional duties often and with great joy and enthusiasm; Kathe Lemon, who set up the steady stream of neighbourly meal deliveries to our door when my wife was rendered immobile by a broken leg in the final weeks of this manuscript's completion; and the guys on the Banjo Hitters Slack, who know who they are and exactly how much of my nonsense they've had to put up with.

The research and field reporting informing this book stretches across nearly twenty years, several other book projects, and numerous grants and fellowships, and was enhanced by the expertise and generosity of hundreds of colleagues around the world. Regarding the new research and reporting for this book, I am especially thankful to: the German government and the Deutsche Gesellschaft für Internationale Zusammenarbeit for their gracious assistance with my media fellowship at the 2019 Berlin Energy Transition Dialogue; my colleagues at the *Walrus* magazine and the Walrus Talks lecture series for facilitating my attendance at GLOBE 2016 and 2020; and energy transition colleagues Merran Smith, Linda Coady, Dan Woynillowicz, Ed Whittingham, Philippe Dunsky, Zoe Caron, Gerald Butts, Mikael Colville-Andersen, the Smart Prosperity Institute, the Canadian Institute for Climate Choices, Clean Energy Canada and the Pembina Institute. My gratitude as well to James Glave and Colleen Giroux-Schmidt for reviewing portions of the draft manuscript and offering timely factchecking help.

It has been a great joy to reunite for this book with the brilliant publishing team at Penguin Random House Canada, particularly Anne Collins, Matthew Sibiga, my editor, Craig Pyette, and publisher, Sue Kuruvilla. I'm thankful as well for Leah Springate's striking design work, the careful copy-editing eye of Gillian Watts, and the watchful proofreading of Tim Hilts.

Finally, I remain more grateful than I could ever adequately express for the love and support of my wife, Ashley Bristowe, and my children, Sloan and Alexander. Whatever optimism I have for this world and our collective future is built on a foundation of the joy they bring me every day.

NOTES

In the interest of brevity, I have not listed the source for every single statistic and detail found in these pages, the vast majority of which are readily located by Google search. Instead, I've cited the sources of all substantial quotes, extensive details drawn from beyond my own reporting and significant arguments that are not mine. Any errors of fact in this book are entirely mine.

1.1 . . . A Better World Waits

1.2 . . . Doom's Limits

1.3 . . . Embrace Dark Euphoria
A full transcript of Bruce Sterling's keynote speech at Reboot 11 on 25 June 2009 can be found at Wired.com (https://www.wired.com/2011/02 /transcript-of-reboot-11-speech-by-bruce-sterling-25-6-2009/).

1.4 . . . An Age of Offhand Miracles
My reporting on EVS-17 in Montreal appeared in the 30 October 2000 issue of *Time* magazine (Canadian edition).

In this section and throughout the book, my primary sources for energy statistics are the International Energy Agency (IEA), the International Renewable Energy Agency (IRENA), and Our World in Data (ourworldindata.org), which relies primarily on the annual *BP Statistical Review of World Energy*.

1.5 . . . Less Bad and Much Better
Estimates on the average speed of pre-industrial modes of travel are from Vaclav Smil, *Energy and Civilization: A History* (Cambridge, MA: The MIT

Press, 2017), 178-188; for details on the rise of the steam locomotive, steam engine and internal combustion engine, see *Ibid.*, 235-50.

1.6 . . . A Much Better World

The Chinese high-speed rail photo described in this section accompanied the article "Eight charts show how 'aggressive' railway expansion could cut emissions" (Jocelyn Timperley, *Carbon Brief*, 30 January 2019): https://www.carbonbrief.org/eight-charts-show-how-aggressive-railway -expansion-could-cut-emissions.

Figures on the size and growth of Chinese and global high-speed rail networks are drawn from a chart at Statistia.com (https://www.statista.com/chart/17093 /miles-of-high-speed-rail-track-in-operation-by-country/), which draws from data provided by the International Union of Railways (UIC).

Chinese HSR growth figures are from *China Daily* (https://www.chinadaily .com.cn/a/202008/13/WS5f34ddfaa3108348172601do.html) and the Trapeze Group (https://www.trapezegroup.com/blog-entry/global-rail-investment -on-the-rise-but-the-us-lags-behind-heres-why).

The figure on the growth of electric bus fleets in China is from Bloomberg News (Brian Eckhouse, "The U.S. Has a Fleet of 300 Electric Buses. China Has 421,000," 15 May 2019): https://www.bloomberg.com/news/articles/2019 -05-15/in-shift-to-electric-bus-it-s-china-ahead-of-u-s-421-000-to-300.

Definitions of "suburban" (vs. urban and/or exurban) are far from standardized. For the ballpark numbers in this section, I used figures from Pew Research Center (https://www.pewresearch.org/social-trends/2018/05 /22/demographic-and-economic-trends-in-urban-suburban-and-rural -communities/) and Bloomberg (https://www.bloomberg.com/news /articles/2018-11-14/u-s-is-majority-suburban-but-doesn-t-define-suburb).

As an Aside: Timelines

See Smil, *Energy and Civilization*: for "dependence" quote, p. 440; for overview of energy transition timelines, 53-4, 153-55, 229-34, 241-50, 265-8, 315-7, and *passim*; for "In 1800..." quote, 267.

2.0 . . . The Long, Looping "Less Bad" Learning Curve

2.1 . . . A Quarter Century of Reckoning with a 200-Year Mess

Details on the process to identify and define the Anthropocene Epoch are drawn from the proceedings of the Subcommission on Quaternary Stratigraphy's Working Group on the "Anthropocene" (http://quaternary.stratigraphy.org/working-groups/anthropocene/), as well as articles published in the journal *Nature* (Meera Subramanian, "Anthropocene Now," *Nature,* 21 May 2019: https://www.nature.com/articles/d41586-019-01641-5) and *The Guardian* (Damian Carrington, "The Anthropocene Epoch: scientists declare dawn of human-influenced age," 29 August 2016: https://www.theguardian.com/environment/2016/aug/29/declare-anthropocene-epoch-experts-urge-geological-congress-human-impact-earth).

2.2 . . . The View from Svaneke

The backstory of Denmark's leadership role in the first phase of the global energy transition is based on my own field reporting since 2005, some of which previously appeared in *The Geography of Hope* (Random House Canada, 2007) and *The Leap* (Random House Canada, 2011).

Background and reports on the EcoGrid project on Bornholm can be found at the project website: http://www.eu-ecogrid.net/.

2.3 . . . On Plausible Optimism

"Energy shock" quote is from an editorial in *The Economist*: "Is it the end of the oil age?," 19 September 2020 (https://www.economist.com/leaders/2020/09/17/is-it-the-end-of-the-oil-age).

2.4 . . . It's Okay To Not Be Okay About It

Britt Wray quote is taken from Susan Shain, "Got Climate Anxiety? These People Are Doing Something About It," *New York Times,* 4 February 2021 (https://www.nytimes.com/2021/02/04/climate/climate-anxiety-stress.html).

Details of Charlie Veron's discovery of the climate crisis in the world's coral reefs are drawn from J.E.N. Veron, *A Reef in Time: The Great Barrier Reef from Beginning to End* (Cambridge, MA: Harvard University Press, 2008).

The news item quoted at length that first alerted me to Veron's work is Jo Chandler, "Scientist's oceanic plea warns of imminent reef eulogy," *The Age* (Melbourne), 7 June 2008 (https://www.theage.com.au/national/scientists-oceanic-plea-warns-of-imminent-reef-eulogy-20080606-2myo.html).

For the *Sydney Morning Herald*'s full report on the decline of the Great Barrier Reef, see Nick O'Malley and Mike Foley, "Barrier reef doomed as up to 99% of coral at risk, report finds," *Sydney Morning Herald*, 1 April 2021 (https://www.smh.com.au/environment/climate-change/barrier-reef-doomed -as-up-to-99-percent-of-coral-at-risk-report-finds-20210331-p57fng.html); online version altered the headline from the print version (image of print version can be found online at https://twitter.com/ukpapers/status /1377406614592839680).

The details of Joanie Kleypas's first discovery of the implications of ocean acidification were reported by the Environmental News Network, among others (see Daniel Glick, "Saving the Oceans: 'Mission Possible,' ENN, 25 February 2009: https://www.enn.com/articles/39367-saving-the-oceans- -mission-possible). I also heard firsthand recollections of the fateful conference from attendees at the Third Symposium on The Ocean in a High-CO^2 World in Monterey, California, in September 2012.

For the quote from Veron's autobiography, see Charlie Veron, *A Life Underwater* (Sydney: Viking, 2017), 296.

2.5 . . . The UN Won't Make It Okay (and That's Okay)
A full transcript of Greta Thunberg's speech to the United Nations Climate Action Summit on 23 September 2019, can be found at NPR: https://www.npr .org/2019/09/23/763452863/transcript-greta-thunbergs-speech-at-the-u-n -climate-action-summit.

Roberts quotes are both from David Roberts, "'Environmentalism' can never address climate change," *Grist*, 10 August 2010 (https://grist.org/article/2010 -08-09-environmentalism-can-never-address-climate-change/).

Marshall quote is from George Marshall, *Don't Even Think About It: Why Our Brains Are Wired to Ignore Climate Change* (New York: Bloomsbury, 2015), 223.

2.7 . . . War Footing and Magical Thinking
Climate Mobilization quote is taken from the "About" page of the organization's website (https://www.theclimatemobilization.org/about/).

Ed Miliband's "war footing" statement was widely reported, for example by Rob Merrick, "UK must fight climate change on 'war footing' like defeat of Nazis, Theresa May told," *The Independent* (UK), 29 April 2019

(https://www.independent.co.uk/news/uk/politics/uk-climate-change
-theresa-may-environment-protest-lucas-ed-miliband-second-world
-war-nazis-a8891801.html).

Quotes and other details from David Wallace-Wells' keynote speech at
Globe 2020 are from my own notes. Some quotes verified or transcribed in
full from a video of the plenary session posted at YouTube (https://www
.youtube.com/watch?v=76OT5GcvM2g).

2.8 . . . The View from Berlin
Details on the early years and growth of the *Energiewende* in Germany
are drawn from my own reporting, which can be found in greater detail
in *The Leap*.

All details and quotes from officials and speakers at the 2019 Berlin Energy
Transition Dialogue are from my notes and transcriptions of my recordings.

As an Aside: On Forecasting
Merkel's "unrealistic" quote was reported in "Merkel pledges German
energy policy changes," *Environment Daily*, 6 September 2005, and later
reprinted in Wolfgang Gründinger, *Drivers of Energy Transition* (Berlin:
Springer VS, 2016), p. 296. Here is the full quote, as translated from the
German by Gründinger: "It is hardly realistic to raise the share of renewables in energy consumption to 20 percent. I believe that it is unrealistic to
expect that renewable energies can close a gap that would be opened by the
early shutdown of nuclear power."

2.9 . . . Epiphany at the Café Einstein
Solyndra quote is from "The Solyndra 'Panic'," *New York Times*,
24 September 2011.

All Hermann Scheer quotes and *Energiewende* detail are from a transcription of my interview with Scheer at the Café Einstein in Berlin on
8 September 2009; some additional background drawn from a transcription of my phone interview with Scheer on 28 May 2008.

Tobias Homann quote from a transcription of my interview with Homann
and Robert Schied at the office of Germany Trade and Investment in Berlin,
7 September 2009.

For details on German public support of the *Energiewende*, see Julian Wettengel, "Polls reveal citizens' support for climate action and energy transition," *Clean Energy Wire*, 14 December 2021 (https://www.cleanenergywire .org/factsheets/polls-reveal-citizens-support-energiewende); includes recap of German public responses from 2012 to 2021 to the polling question "Increased use and expansion of renewable energy is…" Results consistently show more than 90 percent of Germans responding "important" or "extremely important" from 2012 onward (with a dip into the mid-80s during the pandemic years).

2.10 . . . Political Will Is Not the Easy Part
Ramez Naam tweet was posted to the @ramez account on Twitter on 9 February 2019 (https://twitter.com/ramez/status/1094132711126487040).

Details and quotes from Hans-Josef Fell's presentation at the 2019 Berlin Energy Transition Dialogue are from my notes.

David Roberts quote is from Roberts, "The sad truth about our boldest climate target," *Vox*, 3 January 2020 (https://www.vox.com/energy-and -environment/2020/1/3/21045263/climate-change-1-5-degrees-celsius -target-ipcc).

2.11 . . . Climate Politics 101
Roberts quote is from "What is 'political will' anyway? Scholars take a whack at defining it," *Vox*, 24 December 2017 (https://www.vox.com/2016 /2/17/11030876/political-will-definition).

2.12 . . . The Highly Qualified, Necessarily Compromised Thrill of Climate Victory
The Canadian Press coverage of Trudeau's speech can be found at a number of CP subscriber news outlets, including Bruce Cheadle, "Don't pit pipelines against wind turbines, PM says as fed-prov meetings open," *Thunder Bay Chronicle-Journal*, 2 March 2016 (https://www.chroniclejournal.com /news/national/clean-economy-can-be-a-win-for-canada-if-provinces /article_a6903462-9ca4-518f-b13c-87a0d2bcffd0.html).

Mark Jaccard quote is from Jaccard, "Finally Canada is a global example for climate action," *The Globe and Mail*, 15 April 2019 (https://www.theglobeandmail .com/opinion/article-finally-canada-is-global-example-for-climate-action/).

For an example of the "You bought a pipeline" attacks on Trudeau, see Mel Woods, "NDP drops the mic with response to Liberal climate plan," *Huffington Post*, 24 September 2019 (https://www.huffpost.com/archive/ca /entry/ndp-trudeau-climate-pipeline_ca_5d8a684de4b066c9cda59728); includes full text of a federal NDP campaign press release reading, in its entirety, "You. Bought. A. Pipeline."

Gerald Butts quote is from a post to @gmbutts on Twitter, 30 November 2016 (https://twitter.com/gmbutts/status/803971163626106880).

2.13 . . . Memes Are Not Enough

For Truthdig's report on the "100 entities" report, see Lee Camp, "Just 100 companies will sign humanity's death warrant," 15 October 2019 (https://www.truthdig.com/articles/just-100-companies-will-sign -humanitys-death-sentence/).

For further details on the "100 entities" report, see *The Carbon Majors Database: CDP Carbon Majors Report 2017* (CDP Worldwide, 2017, https://cdn.cdp.net/cdp-production/cms/reports/documents/000/002/327 /original/Carbon-Majors-Report-2017.pdf?1501833772).

Peter Dorman quotes are from Dorman, "The Climate Movement Needs to Get Radical, but What Does that Mean?," Nonsite.org, 16 May 2016 (https://nonsite.org/the-climate-movement-needs-to-get-radical-but -what-does-that-mean/).

As an Aside: The Denial Industrial Complex

For an overview of the *Merchants of Doubt* book and documentary film and its findings, see Phoebe Keane, "How the oil industry made us doubt climate change," *BBC News*, 20 September 2020 (https://www.bbc.com/news/stories -53640382); for further detail see also Tik Root et al., "Following the money that undermines climate science," *New York Times*, 10 July 2019 (https://www .nytimes.com/2019/07/10/climate/nyt-climate-newsletter-cei.html), Mark Maslin, "The five corrupt pillars of climate change denial," *The Conversation*, 28 November 2019 (https://theconversation.com/the-five-corrupt-pillars-of -climate-change-denial-122893), and Elliott Negin, "ExxonMobil claims shift on climate but continues to fund climate deniers," *Union of Concerned Scientists* blog, 22 October 2020 (https://blog.ucsusa.org/elliott-negin /exxonmobil-claims-shift-on-climate-continues-to-fund-climate-deniers/).

2.14 . . . Building Codes and the Necessary Process Grind

For an overview of the Saskatachewan Conservation Houe and its connection to the Passivhaus movement, see Mike Reynolds, "Saskatchewan: The birthplace of passive house and passive solar home design," *Ecohome*, 15 October 2013 (https://www.ecohome.net/guides/1422/passive-house-saskatchewan-the -birthplace-of-high-performance-buildings-and-passive-solar-home-design/).

For an overview and history of BC Energy Step Code, see James Glave and Robyn Wark, *Lessons from the BC Energy Step Code* (BC Hydro/Natural Resources Canada publication, June 2019; https://www2.gov.bc.ca/assets /gov/farming-natural-resources-and-industry/construction-industry /building-codes-and-standards/reports/bcenergystepcode_lessons_learned _final.pdf).

2.15 . . . The Science of Not Listening to the Science

Joseph Heath book review quote is from Heath, "Naomi Klein postscript No. 2," *In Due Course* blog, 12 April 2015 (http://induecourse.ca/naomi-klein -postscript-no-2/)

Oliver Morton quote is from Morton, *The Planet Remade: How Geoengineering Could Change the World* (Princeton University Press, 2015), 376.

For an example of Katherine Hayhoe's use of the ten-word summary of the climate crisis, see her post to @KHayhoe on Twitter, 20 September 2019 (https://twitter.com/khayhoe/status/1175038922373640192); as noted, I heard the same firsthand at presentations Hayhoe gave at Globe 2020 and at a community climate event in Calgary in March 2018.

For Hayhoe quote, see Hayhoe, "When Facts are not Enough," *Science*, 1 June 2018 (https://www.science.org/doi/10.1126/science.aau2565).

For Marshall quote, see Marshall, *Don't Even Think About It*, 26-7.

For Kolbert analysis and quotes, see Elizabeth Kolbert, "Why Facts Don't Change Our Minds," *The New Yorker*, 19 February 2017 (https://www .newyorker.com/magazine/2017/02/27/why-facts-dont-change-our-minds).

2.16 . . . Hearts and Minds

Kahneman quote is from Marshall, *Don't Even Think About It*, 56.

Overview of prospect theory and all other details regarding Kahneman and Tversky's work from Daniel Kahneman, *Thinking, Fast and Slow* (Toronto: Doubleday Canada, 2011). For prospect theory quote and explanation, see 282-87. See also 119-125, 137-40, 199-207.

For Dan Ariely's work on anchoring effects, see Dan Ariely, *Predictably Irrational: The Hidden Forces That Shape Our Decisions* (revised and expanded edition; New York: HarperCollins, 2009), 25-36. See also "Dissecting people's 'predictably irrational' behaviour," NPR radio, 21 February 2008 (https://www.npr.org/transcripts/19231906).

For a succinct overview of the endowment effect and Richard Thaler's seminal work on it, see "Why do we value items more if they belong to us?" at *The Decision Lab* blog (https://thedecisionlab.com/biases/endowment-effect/).

For Daniel Gilbert's argument about the four critical aspects of climate change, see "Humans wired to respond to short-term problems," NPR radio, 3 July 2006 (https://www.npr.org/templates/story/story.php?storyId=5530483).

3.0 . . . The Much Better Decade

3.1 . . . The View from Quarantine
For details of the Tesla home in Kauai, see Carrie Coolidge, "Tesla owners will love this stunning home on Kauai," *Forbes*, 14 September 2018.

3.2 . . . Reality Check
For Hawken praise, see for example Norbert Senf, "Paul Hawken and the Ecology of Commerce," *MHA News*, 1994 (https://www.mha-net.org/docs /hawken.htm).

For a list of Project Drawdown's emissions reduction solutions and other details of the project, see the organization's website (https://drawdown.org /solutions/table-of-solutions).

Hawken quotes at Eco-City Summit are from my transcription, full video of Hawken's keynote can be found on YouTube (https://www.youtube.com /watch?v=kkEdl89-B9U).

For Joseph Heath's explanation of "Hobbes' difficult idea," see Heath, "Hobbes' difficult idea," *In Due Course* blog, 15 December 2014 (http://induecourse.ca/hobbess-difficult-idea/). My summary of the idea also draws on my interview with Heath by phone on 20 June 2019, from which the long quotes here are taken.

3.3 . . . Pledge Drives

For Greenpeace statement on Shell's net zero pledge, see Ron Boussa and Shadia Nasralla, "Shell sets emissions ambition of net zero by 2050, with customer help," Reuters, 16 April 2020 (https://www.reuters.com/article /us-shell-emissions-idUSKCN21Y0MW).

The Canadian Institute for Climate Choices report *Canada's Net Zero Future* (February 2021) is available for download at the Institute's website (https://climatechoices.ca/reports/canadas-net-zero-future/).

3.4 . . . Objects in Motion

3.5 . . . The Green Marshall Plan

My overview of the history of the Marshall Plan is based primarily on Greg Behrman, *The Most Noble Adventure: The Marshall Plan and How America Helped Rebuild Europe* (Free Press, 2008), especially p. 56-61, 112-120, 134-36, 188-97, 286-294. Long Truman quote is from his speech to Congress on 17 November 1947, see 134, short quote from a statement on 3 April, 1948, see 163.

My synopsis of the rise of solar power is based on my own reporting, including quote from Tobias Homann (previously cited in Section 2.9).

Siemens CEO quote is from my BETD 2019 notes. For an example of the company's description of the scale of the Crossrail project, see the Siemens report *Going Underground* (https://assets.new.siemens.com/siemens/assets /api/uuid:abf728ef-e796-4e13-b90b-36f348e79eaf/presentation-going -underground-e.pdf).

Laurence Tubiana quote is from a post to the account @LaurenceTubiana, made on 12 December 2020 (https://twitter.com/LaurenceTubiana/status /1337724793228046337).

3.6 . . . California Dreaming

James Howard Kunstler quote is from Kunstler, *The Geography of Nowhere: The Rise and Decline of America's Man-made Landscape* (Free Press, 1994);

Kunstler has repeated the line often in many forums, including the book's description at his own website (https://kunstler.com/other-stuff/geography -of-nowhere-e-book-out/).

Synopsis of California's role in framing the suburban dream is from Kirse Granat May, *Golden State, Golden Youth: The California Image in Popular Culture, 1955-1966* (University of North Carolina Press, 2002), *passim*. For quote, see 4.

For Musk quote, see Brett Arends, "Tesla's solar-roof sales will grow 'like kelp on steroids,' Musk vows," *MarketWatch*, 30 October 2019 (https://www.marketwatch.com/story/teslas-solar-roof-sales-will-grow -like-kelp-on-steroids-musk-vows-2019-10-27).

3.7 . . . The Good Life Rebooted

3.8 . . . Life in Energy-Transition Disneyland
Much of the detail on Bornholm in this section is drawn from my field research and reporting in 2013 and 2019. As per earlier note, further back-ground and reports on the EcoGrid project can be found at the project website: http://www.eu-ecogrid.net/.

For an overview of Hawaii's energy transition, see Noelle Swan and Nathan Eagle, "How Hawaii has built momentum to become a renew-able energy leader," *GreenBiz*, 26 September 2019 (https://www.greenbiz .com/article/how-hawaii-has-built-momentum-become-renewable-energy-leader).

As an Aside: New Wonders of the World
For full details on the SiemensGamesa 14-222 wind turbine, see the com-pany's website: https://www.siemensgamesa.com/products-and-services /offshore/wind-turbine-sg-14-222-dd.

3.9 . . . The Better Urban Life
The general details on the emergence of Copenhagen as an urban cycling model are from my own reporting there in 2005, 2009, 2013 and 2019, including extensive interviews with Mikael Colville-Anderson, Jan Gehl, and other local cycling and planning officials.

For details on the survey of Copenhagen cycling attitudes, see Julia Day and David Sim, "Copenhagen's Lessons for the 'Green Wave,'" *Streetsblog NYC*, 22 November 2019 (https://nyc.streetsblog.org/2019/11/22/op-ed-copenhagens-lessons-for-the-green-wave/). See also Erik Kirschbaum, "Copenhagen has taken bicycle commuting to a whole new level," *Los Angeles Times*, 7 August 2019 (https://www.latimes.com/world-nation/story/2019-08-07/copenhagen-has-taken-bicycle-commuting-to-a-new-level).

3.10 . . . Much Better Blocks
Quotes from Mike Lydon and background detail on tactical urbanism are drawn primarily from my interview with Lydon in July 2012 and subsequent interviews and informal conversations with him and other tactical urbanists. Final Lydon quote ("No city will build a bridge…") taken from Kim A. O'Connell, "Newest Urbanism," *AIA Architect*, 9 July 2013 (https://www.architectmagazine.com/aia-architect/aiafuture/newest-urbanism-1_0).

Jason Roberts quote and background detail on first Better Block from Lisa Gray, "Building a Better Block," *Houston Chronicle*, 28 June 2010 (https://www.chron.com/entertainment/article/Gray-Building-a-better-block-1711370.php); for the Andrew Howard quote and further background, see Mike Lydon and Anthony Garcia, "How One Weekend in Dallas Sparked a Movement for Urban Change," *Next City*, 20 April 2015 (https://nextcity.org/features/how-one-weekend-in-dallas-sparked-a-movement-for-urban-change) and the Better Block Foundation website (https://www.betterblock.org/about).

3.11 . . . Ode to Unsung Climate Heroes, Part 1: Density
For details of the Sightline Insitute's density study, see Michael Andersen, "A Duplex, a Triplex and a Fourplex can cut a block's carbon impact 20%," *Sightline Institute blog*, 7 June 2019 (https://www.sightline.org/2019/06/07/a-duplex-a-triplex-and-a-fourplex-can-cut-a-blocks-carbon-impact-20/).

Residential zoning statistics are drawn from Emily Badger and Quoctrong Bui, "Cities start to question an American ideal: A house with a yard on every lot," *New York Times*, 18 June 2019 (https://www.nytimes.com/interactive/2019/06/18/upshot/cities-across-america-question-single-family-zoning.html).

3.12 . . . Ode to Unsung Climate Heroes, Part 2: Efficiency

My description of Unilever-Haus in Hamburg is based on my notes from a tour of the building in 2011. For further background, see Andrew Michler, "Unilever's energy efficient office is one of the greenest in Europe," *Inhabitat*, 7 January 2011 (https://inhabitat.com/hamburgs-unilever-headquarters -is-cities-second-skin/).

For details on Robert Cialdini's hotel towel and energy bill experiments, see Noah Goldstein, "Changing Minds and Changing Towels," *Psychology Today*, 23 August 2008 (https://www.psychologytoday.com/ca/blog/yes /200808/changing-minds-and-changing-towels) and Mark Joseph Stern, "A Little Guilt, a Lot of Energy Savings," *Slate*, 1 March 2013.

For background on the function and efficiency gains of heat pumps, see Ula Chrobak, "How heat pumps can help fight global warming," *Popular Science*, 3 March 2020 (https://www.popsci.com/story/environment/heat- pumps-emissions-climate-change/); Rachel Golden and Cara Bottorf, "Heat pumps slow climate change in every corner of the country," *Sierra Club* blog, 23 April 2020 (https://www.sierraclub.org/articles/2020/04/new -analysis-heat-pumps-slow-climate-change-every-corner-country); and Prachi Patel, "Heat pumps could shrink the carbon footprint of buildings," *IEEE Spectrum*, 19 September 2019 (https://spectrum.ieee.org/heat-pumps -could-shrink-the-carbon-footprint-of-buildings).

3.13... Electric Age

Details of the development of British Columbia's energy step code are drawn from my own conversations with James Glave, as well as Glave, *Lessons from the BC Energy Step Code* (Natural Resources Canada/BC Hydro, 2019; https://www2.gov.bc.ca/assets/gov/farming-natural-resources -and-industry/construction-industry/building-codes-and-standards /reports/bcenergystepcode_lessons_learned_final.pdf).

For background on Costa Rica's clean energy, climate and transportation planning, see Paul Rubio, "Reaching for a zero-emission goal," *New York Times*, 21 September 2018 (https://www.nytimes.com/2018/09/21/climate /costa-rica-zero-carbon-neutral.html); "Solving Costa Rica's traffic and pollution problem," *Tico Times*, 16 July 2018 (https://ticotimes.net/2018/07/16 /solving-costa-ricas-traffic-and-pollution-problem); Sebastian Rodriguez, "Costa Rica's president-elect promises zero-carbon transport," *Christian Science Monitor*, 1 May 2018 (https://www.csmonitor.com/World/Americas

/2018/0501/Costa-Rica-s-president-elect-promises-zero-carbon-transport);
Lindsay Fendt, "All that glitters is not green," *The Guardian,* 5 January 2017
(https://www.theguardian.com/world/2017/jan/05/costa-rica-renewable
-energy-oil-cars); Diane Toomey, "How Costa Rica is Moving Toward a Green
Economy," *Yale Environment 360,* 5 January 2017 (https://e360.yale.edu
/features/how_costa_rica_is_moving_toward_a_green_economy_renewable
_energy); and "Costa Rica's urban transit plan going nowhere," *Tico Times,*
22 July 2011 (https://ticotimes.net/2011/07/22/costa-rica-s-urban-transit
-plan-going-nowhere).

Noah Smith quotes taken from Smith, "We will not ban cars," *Noahpinion*
Substack blog, 9 March 2021 (https://noahpinion.substack.com/p/we-will
-not-ban-cars).

3.14 . . . The New Industrial Age
Washington Post quote about Elysis is from Chris Mooney, "This could be
the biggest advance in aluminum production in 130 years," *Washington Post,*
14 May 2008 (https://www.washingtonpost.com/news/energy-environment
/wp/2018/05/14/this-could-be-the-biggest-advance-in-aluminum-production
-in-130-years/).

Marc Gunther quote is from Gunther, "Walmart is slapping itself on the back
for sustainability but it still has a way to go," *The Guardian,* 18 November 2015
(https://www.theguardian.com/sustainable-business/2015/nov/18/walmart
-climate-change-carbon-emissions-renewabe-energy-environment).

Walmart announced its "regenerative company" goal in a press release,
"Walmart sets goal to become regenerative company," 21 September 2020,
available at the company website (https://corporate.walmart.com/newsroom
/2020/09/21/walmart-sets-goal-to-become-a-regenerative-company).

3.15 . . . Power and Justice
For details on the Athabasca Chipewyan First Nation's solar projects, see
Jordan Omstead, "Indigenous-owned solar farm opens in remote northern
community," *CBC News,* 19 November 2020 (https://www.cbc.ca/news
/canada/edmonton/indigenous-owned-solar-farm-fort-chip-1.5807721) and
"ACFN announces major investment in Alberta clean electricity with
Concord Pacific" (press release), 4 May 2021 (https://gpenergyanalytics.ca
/acfn-green-energy/).

3.16 . . . Levers and Incentives

The IEA quote is from the IEA's *World Energy Outlook* report for 2020, quoted in Simon Evans, "Solar is now 'cheapest electricity in history,'" *Carbon Brief*, 13 October 2020.

"Climate club" overview and William Nordhaus quotes are from Nordhaus, "A New Solution: The Climate Club," *New York Review of Books*, 4 June 2015. See also Nordhaus, "The Climate Club: How to fix a failing global effort," *Foreign Affairs*, May/June 2020.

3.17 . . . The Humming Twenties

BNP Paribas quote is drawn from the 2019 report *Wells, Wires and Wheels*, quoted in Henry Edwardes Evans, "Wind/solar twinned with EVs 'pose existential threat to gasoline': BNP Paribas," *S&P Global Platts*, 6 August 2019 (https://www.spglobal.com/platts/en/market-insights/latest-news /oil/080619-wind-solar-twinned-with-evs-pose-existential-threat-to -gasoline-bnp-paribas).

For Bloomberg quote, see Lynn Doan et al., "What's Behind the World's Biggest Climate Victory? Capitalism," *Bloomberg News*, 15 September 2019 (https://www.bloomberg.com/graphics/2019-can-renewable-energy-power -the-world/)

IEA "new normal" quote is drawn from *Renewable Energy Market Update 2021* (IEA, 2021; https://www.iea.org/reports/renewable-energy-market -update-2021).

McKinsey quote is drawn from *Global Energy Perspective 2019* (McKinsey Solutions 2019; https://www.mckinsey.com/~/media/McKinsey/Industries /Oil%20and%20Gas/Our%20Insights/Global%20Energy%20Perspective %202019/McKinsey-Energy-Insights-Global-Energy-Perspective-2019 _Reference-Case-Summary.ashx).

New Jersey offshore wind quote is from Tim Sullivan of the New Jersey Economic Development Authority, quoted in Dino Grandoni, "New Jersey aims to lead nation in offshore wind. So it's building the biggest turbine port in the country," *Washington Post*, 16 June 2020 (https://www.washingtonpost .com/climate-solutions/2020/06/16/new-jersey-aims-lead-nation-offshore -wind-so-its-building-biggest-turbine-port-country/).

3.18 . . . Adventures in Necessary Sci-Fi

For details on CICC modelling of Canada's net zero pathway, see *Canada's Net Zero* (February 2021; https://climatechoices.ca/reports/canadas-net-zero-future/).

The IPCC's findings on the necessity of carbon removal are in its Fifth Assessment Report (*AR5 Synthesis Report: Climate Change 2014*; https://www.ipcc.ch/report/ar5/syr/); for a summary, see the "carbon removal factsheet" at the American University Institute for Carbon Removal Law and Policy website (https://www.american.edu/sis/centers/carbon-removal/copy-of-copy-of-fact-carbon-removal.cfm).

For Liebreich quotes and analysis, see Michael Liebreich, "Beyond Three Thirds: The Road to Deep Decarbonization," *BNEF* blog, 13 March 2018 (https://about.bnef.com/blog/liebreich-beyond-three-thirds-road-deep-decarbonization/).

Leibreich quote on Eavor's potential is from Leigh Collins, "Unlimited, on-demand energy anywhere in the world—is Eavor-Loop climate change's holy grail?" *Recharge News*, 27 October 2020 (https://www.rechargenews.com/transition/unlimited-on-demand-renewable-energy-anywhere-in-the-world-is-eavor-loop-climate-changes-holy-grail-/2-1-901385).

For Morton quote see Morton, *The Planet Remade*, 166-67

INDEX

CHRIS TURNER is a three-time nominee and one-time winner of the National Business Book Award, and a finalist for the Governor General's Literary Award for Nonfiction (for *The Geography of Hope*). He has long been one of Canada's leading voices on climate solutions and the global energy transition. His feature writing has earned ten National Magazine Awards and he is the author of five books on technology, energy and climate. He lives in Calgary with his wife, the author Ashley Bristowe, and their two children.